ANTENNA AND EM MODELING WITH MATLAB

ANTENNA AND EM MODELING WITH MATLAB

Sergey N. Makarov

A JOHN WILEY & SONS, INC., PUBLICATION

This book is printed on acid-free paper. ∞

Copyright © 2002 by John Wiley and Sons, Inc., New York. All rights reserved.

Published simultaneously in Canada.

No part of this publication may be reproduced, stored in a retrieval system or transmitted in any form or by any means, electronic, mechanical, photocopying, recording, scanning or otherwise, except as permitted under Sections 107 or 108 of the 1976 United States Copyright Act, without either the prior written permission of the Publisher, or authorization through payment of the appropriate per-copy fee to the Copyright Clearance Center, 222 Rosewood Drive, Danvers, MA 01923, (978) 750-8400, fax (978) 750-4744. Requests to the Publisher for permission should be addressed to the Permissions Department, John Wiley & Sons, Inc., 605 Third Avenue, New York, NY 10158-0012, (212) 850-6011, fax (212) 850-6008, E-Mail: PERMREQ@WILEY.COM.

For ordering and customer service, call 1-800-CALL-WILEY.

Library of Congress Cataloging-in-Publication Data is available.

ISBN: 0-471-21876-6

Printed in the United States of America.

10 9 8 7 6 5 4

To my parents

CONTENTS

Preface   xiii

1   Introduction   1

1.1.   Matlab / 1
1.2.   Antenna Theory / 2
1.3.   Matlab Codes / 3
1.4.   Antenna Structures / 3
1.5.   Method of Analysis and Impedance Matrix / 5
1.6.   Wire and Patch Antennas / 7
1.7.   Matlab Loops and Antenna Optimization / 7
1.8.   Speed and Maximum Size of the Impedance Matrix / 8
1.9.   Outline of Chapters / 8
       References / 10

2   Receiving Antenna: The Scattering Algorithm   11

2.1.   Introduction / 11
2.2.   Code Sequence / 13
2.3.   Creating the Antenna's Structure / 13
2.4.   RWG Edge Elements / 16
2.5.   Impedance Matrix / 19
2.6.   Moment Equations and Surface Currents / 21
2.7.   Visualization of Surface Currents / 23
2.8.   Induced Electric Current of a Dipole Antenna / 25
2.9.   Induced Electric Current of a Bowtie Antenna / 28
2.10.   Induced Electric Current of a Slot Antenna / 29
2.11.   Using the Matlab Compiler / 33
2.12.   Using Matlab for Linux / 34

2.13. Conclusions / 34
References / 35
Problems / 36

3 Algorithm for Far and Near Fields   39

3.1. Introduction / 39
3.2. Code Sequence / 40
3.3. Radiation of Surface Currents / 42
3.4. Far Field / 44
3.5. Radiated Field at a Point / 44
3.6. Radiation Density/Intensity Distribution / 46
3.7. Antenna Directivity / 47
3.8. Antenna Gain (Ideal Case) / 50
3.9. Antenna's Effective Aperture / 51
3.10. Conclusions / 52
References / 53
Problems / 53

4 Dipole and Monopole Antennas: The Radiation Algorithm   57

4.1. Introduction / 57
4.2. Code Sequence / 58
4.3. Strip Model of a Wire / 60
4.4. Feed Model / 60
4.5. Current Distribution of the Dipole Antenna / 63
4.6. Input Impedance / 65
4.7. Monopole Antenna / 66
4.8. Impedance of the Monopole / 71
4.9. Radiation Intensity, Radiated Power, and Gain / 72
4.10. Radiation Resistance and Delivered Electric Power / 74
4.11. Directivity Patterns / 76
4.12. Receiving Antenna / 79
4.13. Friis Transmission Formula / 81
4.14. Conclusions / 82
References / 83
Problems / 84

5 Loop Antennas   89

5.1. Introduction / 89
5.2. Loop Meshes and the Feeding Edge / 90

5.3. Current Distribution of a Loop Antenna / 91
5.4. Input Impedance of a Small Loop / 95
5.5. Radiation Intensity of a Small Loop / 96
5.6. Radiation Patterns of a Small Loop / 98
5.7. Transition from Small to Large Loop: The Axial Radiator / 100
5.8. Helical Antenna—Normal Mode / 102
5.9. Helical Antenna—Axial Mode / 105
5.10. Conclusions / 109
References / 109
Problems / 110

6 Antenna Arrays: The Parameter Sweep 113

6.1. Introduction / 114
6.2. Array Generators: Linear and Circular Arrays / 115
6.3. Array Terminal Impedance / 118
6.4. Impedance and Radiated Power of Two-Element Array / 120
6.5. How to Organize the Matlab Loop / 122
6.6. Array Network Equations / 123
6.7. Directivity Control / 124
6.8. Broadside Array / 124
6.9. End-Fire Array / 128
6.10. Pattern Multiplication Theorem / 130
6.11. Comparison of Theory and Simulation / 132
6.12. Optimization of End-Fire Array: The Phase Loop / 133
6.13. Hansen-Woodyard Model / 135
6.14. Power Map of End-Fire Array / 137
6.15. Phased (Scanning) Array / 139
6.16. Array of Bowties over Ground Plane / 141
6.17. On the Size of the Impedance Matrix / 144
6.18. Conclusions / 147
References / 147
Problems / 148

7 Broadband Antennas: The Frequency Sweep 151

7.1. Introduction / 151
7.2. Code Sequence / 153
7.3. Antenna Structures Under Study / 155
7.4. Dipole Impedance and Power Resonance / 158
7.5. Dipole Radiated Power, Return Loss, and Gain / 160

- 7.6. Dipole Comparison with NEC Modeling / 162
- 7.7. Matlab Mesh for Bowtie Antenna Using Delaunay / 165
- 7.8. Bowtie Impedance / 166
- 7.9. Bowtie Radiated Power and Gain / 166
- 7.10. Bowtie Radiation Intensity Distribution / 169
- 7.11. Mesh for a Spiral Antenna / 172
- 7.12. Spiral Antenna's Impedance, Power, and Gain / 172
- 7.13. Spiral Antenna's Radiation Intensity Distribution / 176
- 7.14. Multiband Antennas: The Sierpinski Fractal / 179
- 7.15. Sierpinski Fractal's Impedance, Power, and Gain / 181
- 7.16. Conclusions / 184
  References / 186
  Problems / 187

8 Ultra-wideband Communication Antenna: Time Domain Analysis   193

- 8.1. Introduction / 193
- 8.2. Code Sequence / 195
- 8.3. Incident Voltage Pulse / 197
- 8.4. Surface Discretization and Feed Model / 199
- 8.5. Frequency Loop / 200
- 8.6. Surface Current Distribution / 201
- 8.7. Antenna Input Impedance / 201
- 8.8. Antenna Radiation Intensity, Gain / 204
- 8.9. Directivity Patterns / 205
- 8.10. Antenna-to-Free–Space Transfer Function / 207
- 8.11. Antenna-to-Antenna Transfer Function / 209
- 8.12. Discrete Fourier Transform / 211
- 8.13. Received Voltage Pulse / 212
- 8.14. Impedance Mismatch / 213
- 8.15. Voltage Pulse at a Load / 216
- 8.16. Conclusions / 218
  References / 219
  Problems / 220

9 Antenna Loading: Lumped Elements   223

- 9.1. Introduction / 223
- 9.2. Code Sequence / 224

9.3. Lumped Resistor, Inductor, and Capacitor / 224
9.4. Test / 227
9.5. Effects of Resistive and Capacitive Loading / 227
9.6. Conclusions / 228
      References / 231
      Problems / 231

10 Patch Antennas   233

10.1. Introduction / 233
10.2. Code Sequence / 234
10.3. Model of the Probe Feed / 236
10.4. Generation of the Antenna Structure / 237
10.5. Input Impedance, Return Loss, and the Radiation Pattern / 239
10.6. Why Do We Need a Wide Patch? / 241
10.7. A Practical Example / 245
10.8. Dielectric Model / 248
10.9. Accuracy of the Dielectric Model / 249
10.10. Conclusions / 251
       References / 253
       Problems / 254

Appendix A Other Triangular Meshes   259

Appendix B Impedance Matrix Calculation   265

Index   269

PREFACE

This text uses the standard Matlab® package in order to model and optimize radiation and scattering of basic RF and wireless communication antennas and microwave structures. The full-wave solution is given by the method of moments. The antenna structures range from simple dipoles to patch antennas and patch antenna arrays. Some special antenna types, such as fractal antennas are considered as well.

The antenna theory presented serves only as necessary background. Each book chapter has an associated Matlab directory on the CD-ROM, containing the Matlab source codes and the antenna generator codes. The reader is encouraged to further develop, improve, or replace any of these codes for his/her own needs.

The Matlab package supports the moment method in the sense that it already has highly efficient built-in matrix solvers. The impedance matrix itself is created in this text using Rao-Wilton-Glisson (RWG) basis functions, the electric field integral equation, and the feeding edge model.

Another very inviting property of Matlab is the wide variety of two- and three-dimensional visualization tools. These tools make antenna analysis an exciting journey. Matlab 6 Release 12 also has built-in mesh generators that allow one to create complicated antenna structures in a simple and observable way.

For course use, the text is intended primarily in combination with Balanis's book *Antenna Theory*, as a simulation and development tool. Some of the chapters can be omitted without loss of continuity.

I would like to acknowledge the suggestions and constructive criticism of the reviewers for this book. My students Tony Garcia and Anuja Apte provided numerous corrections to the text. I am grateful to the staff of John Wiley & Sons, Inc. for the interest and support in the publication of this text. I am thankful to Dr. James Mink who made this work possible.

Finally, I thank my loving wife Natasha for many good things but for too many to list.

Sergey Makarov
ECE Department
Worcester Polytechnic Institute, MA

1

INTRODUCTION

1.1. Matlab
1.2. Antenna Theory
1.3. Matlab Codes
1.4. Antenna Structures
1.5. Method of Analysis and Impedance Matrix
1.6. Wire and Patch Antennas
1.7. Matlab Loops and Antenna Optimization
1.8. Speed and Maximum Size of the Impedance Matrix
1.9. Outline of Chapters
References

1.1. MATLAB

Standard Matlab® has all the ingredients needed to develop a simple and interactive "antenna toolbox." These include (1) built-in surface and volume mesh generators in two and three dimensions, (2) highly efficient matrix solvers, (3) Fourier analysis tools, and (4) 2D and 3D plots with rotation, zoom, and scroll options.

To date, there are two EM antenna-related packages built on Matlab: The Femlab® software of COMSOL group (Sweden, Finland, United States), and SuperNEC from Poynting Software Ltd., South Africa. In SuperNEC, an executable C++ file is still employed in order to perform the key MoM calculations, whereas Matlab essentially plays the role of a GUI.

Femlab includes a number of finite element modules. Among these is an

"Electromagnetics Module" developed to perform finite element simulations of quasi-static and high-frequency fields in Matlab. The emphasis here is on quasi-statics and microwaves, but not on the antenna analysis. The complete Femlab EM module is also rather expensive.

This text explains how to use the standard Matlab package in order to simulate radiation and scattering of basic RF and wireless communication antennas and microwave structures. Rao-Wilton-Glisson (RWG) basis functions, the electric field integral equation, and the feeding-edge model are the background material of the underlying MoM (method of moments) code.

1.2. ANTENNA THEORY

In writing this book, I attempted to strike a balance in two ways. First, while the book does present Matlab code, it is not a programming manual. Appropriate references are given for those readers not already familiar with Matlab.

Second, while antenna theory is developed consistently, this is not a general text on antennas. In presenting the theory, I took cues from Balanis's famous text, *Antenna Theory* [1]. Numerous examples are given to show how to calculate practically important antenna/target parameters. These parameters include:

- Surface current distribution
- Input impedance and return loss
- Antenna near field
- Antenna far field
- Radiation intensity and radiation density
- Radar cross section
- Directivity pattern, gain and beamwidth
- Antenna transfer function and antenna-to-antenna link
- Time domain antenna characteristics

Many chapters are further organized in such a way as to directly support the Balanis text. In particular,

Chapter 4 Dipole and Monopole Antenna
Chapter 5 Loop Antennas
Chapter 6 Antenna Arrays
Chapter 7 Broadband Antennas
Chapter 10 Patch Antennas

complement Chapters 4, 5, 6, 9, 10, and 14 of Balanis's book.

1.3. MATLAB CODES

Every book chapter has an associated Matlab directory in the CD-ROM, containing the corresponding Matlab source codes and the antenna mesh generator files. These codes are executed sequentially and reflect successive numerical steps of the moment method. Before reading the chapter, the reader may want to run these scripts first. The final script will display the results for the starting example of the corresponding chapter.

The "core" of the book consists of two relatively short (each less than 70 lines) Matlab scripts. The first script `impmet.m` computes the impedance matrix Z using the algorithm developed in [2]. This is the basic and the most important step of the moment method. The impedance matrix allows us to determine electric currents flowing on the antenna surface. Once the impedance matrix is programmed correctly, the rest of the work usually does not constitute any difficulties. The Matlab function `impmet.m` is used in the Matlab directory of every chapter.

The second code `point.m` computes the radiated field of an infinitesimally small electric dipole or a group of dipoles at any point in space. This code helps us to determine near and far field of an antenna, including its radiation patterns and gain. The Matlab function `point.m` is used in the Matlab directory of every chapter except for that of Chapter 2.

Both codes are almost fully vectorized and comparable to the corresponding C++ executable files with respect to speed. The interested reader could try to modify these codes to improve their performance.

The rest of the Matlab scripts are used to support and visualize the input and output data for an antenna. Codes, whose name starts with `rwg`, e.g., `rwg1.m`, `rwg2.m`, and `rwg3.m`, are responsible for operations with the antenna structure. Codes, whose names start with `efield`, such as `efield1.m`, `efield2.m`, and `efield3.m`, are responsible for near- and far-field antenna parameters. The reader is encouraged to further develop, improve, or replace any of these codes.

1.4. ANTENNA STRUCTURES

An important problem is how to create an antenna structure. Matlab provides several ways of doing so. One way is to use the built-in mesh generator of the Matlab PDE toolbox. This mesh generator creates planar structures of any (intercepting) rectangles, polygons, and circles, using the convenient graphical user interface (GUI). To extend the design to a 3D structure, it is usually enough to write a short Matlab script involving the z-coordinate dependency.

Another way is to identify the boundary of the antenna structure analytically. Then Delaunay triangulation is applied to that structure, using Matlab function `delaunay`. To approach 3D structures, function `delaunay3` (3D

4 INTRODUCTION

Table 1.1. Antenna Mesh Generators Available from Subdirectory **Mesh** of the Matlab Directory of Each Chapter

	Chapter
Antenna	
Dipole/strip	4, 6
Dipole with reflector(s)	4
Linear/circular array of dipoles	6
Monopole on a finite ground plane	4, 6
Array of monopoles on a finite ground plane	4, 6
Bowtie antenna	6, 7
Array of bowties	6
Loop antenna	5
Helical antenna	5
Spiral antenna	7
Fractal antenna (Sierpinski fractal)	7
Patch antenna on a finite ground plane	10
Patch array on a finite ground plane	10
Structure	
Plate, cube, sphere	Appendix A
Metal meshes (frequency selective surfaces)	10, Appendix A

Delaunay tessellation) may be used. The advantage of this approach is that we do not need the PDE toolbox and can create arbitrary 3D antenna surface and volume meshes. Also the otherwise created meshes have been imported to Matlab in ASCII format.

The library of antenna mesh generators from the text covers about 15 basic antenna shapes to date. Each antenna mesh generator is a Matlab script that outputs the antenna structure. Input parameters (length, width, height, discretization, etc.) are subject to change. Detailed explanations and examples are given in order to help readers to create their own antenna structures. Table 1.1 lists the mesh generator scripts.

The basic shapes can be bent, cloned, and combined with each other using simple Matlab operations, as explained in the main text. In particular, we are able to create such structures as patch arrays with arbitrarily shaped patches on finite ground planes or on finite metal meshes (frequency selective surfaces) including metal lenses.

Some mesh generators (monopole on a finite ground plane or patch antenna/array on a finite ground plane) use Matlab mouse input to enable easy identification of the antenna elements. Figure 1.1 gives an example of

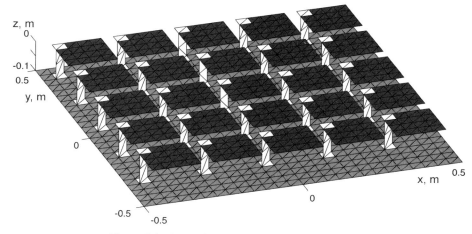

Figure 1.1. Array of patches on a finite ground plane.

the finite patch array created in this way (see Problems 10.9 and 10.10 of Chapter 10).

1.5. METHOD OF ANALYSIS AND IMPEDANCE MATRIX

The method of moments (MoM) used in this text relies on RWG (Rao-Wilton-Glisson) edge elements [2]. First, the surface of a metal antenna under study is divided into separate triangles as shown in Fig. 1.2a. Each pair of triangles, having a common edge, constitutes the corresponding RWG edge element; see Fig. 1.2b. One of the triangles has a plus sign and the other a minus sign. A vector function (or basis function)

$$\mathbf{f}(\mathbf{r}) = \begin{cases} (l/2A^+)\boldsymbol{\rho}^+(\mathbf{r}), & \mathbf{r} \text{ in } T^+ \\ (l/2A^-)\boldsymbol{\rho}^-(\mathbf{r}), & \mathbf{r} \text{ in } T^- \\ 0, & \text{otherwise} \end{cases} \quad (1.1)$$

is assigned to the edge element. Here l is the edge length and $A^\pm$ is the area of triangle $T^\pm$. Vectors $\boldsymbol{\rho}^\pm$ are shown in Fig. 1.2b. Vector $\boldsymbol{\rho}^+$ connects the free vertex of the plus triangle to the observation point $\mathbf{r}$. Vector $\boldsymbol{\rho}^-$ connects the observation point to the free vertex of the minus triangle.

The surface electric current on the antenna surface (a vector) is a sum of the contributions (1.1) over all edge elements, with unknown coefficients.

6 INTRODUCTION

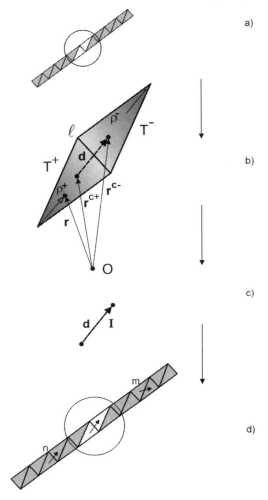

Figure 1.2. Schematic of a RWG edge element and the dipole interpretation.

These coefficients are found from the moment equations discussed in Chapter 2. The moment equations are a linear system of equations with the impedance matrix Z.

The basis function (1.1) of the edge element approximately corresponds to a small but finite electric dipole of length $d = |\mathbf{r}^{c-} - \mathbf{r}^{c+}|$; see Fig. 1.2$b$ and c. Index c denotes the center of triangle $T^{\pm}$. Thus the division of the antenna structure into RWG edge elements approximately corresponds to the division of the antenna current into small "elementary" electric dipoles; see Fig. 1.2d.

In this sense, the impedance matrix Z describes the interaction between different elementary dipoles. If the edge elements m and n are treated as small electric dipoles, the element Z_{mn} of the impedance matrix Z describes the contribution of dipole n (through the radiated field) to the electric current of dipole m, and vice versa. The size of the impedance matrix is equal to the number of the edge elements. This contribution can be either calculated analytically (using the analytical solution for the finite dipole) or by employing the electric field integral equation (EFIE) [2,3].

The RWG edge elements are, however, more advantageous than the simple finite dipoles shown in Fig. 1.2d. In particular, they support a uniform axial electric current along a thin metal strip. The corresponding results are presented in Chapter 4. This circumstance is important for modeling wire antennas using the RWG edge elements [5,6].

1.6. WIRE AND PATCH ANTENNAS

Wire antennas are traditionally studied in terms of a one-dimensional segment model (numerical code NEC). Theory behind NEC requires a special integral equation and a special set of basis functions. Thus there are two different models that we must keep track of: one model for conducting surfaces/patches and another model for wires [4]. More problems arise when wires and surfaces are joined together [4].

To avoid the development and programming of two separate algorithms, the potential of RWG boundary elements for modeling wire antennas is investigated [5,6]. A wire is represented with the use of a thin strip model having one RWG edge element per strip width. The strip width should be four times the wire radius [1]. The results obtained are encouraging. Faithful reproduction of the surface current distribution, input impedance, and gain are observed for dipole and monopole antennas. Therefore both patch and wire antennas will be described in the text using the same basis functions—RWG edge elements [7,8]. This greatly simplifies the underlying math and Matlab source codes for monopole antennas and probe-fed patch antennas.

1.7. MATLAB LOOPS AND ANTENNA OPTIMIZATION

Two Matlab loops are discussed in the text: the loop denoting the route of an antenna parameter (antenna spacing, phase of the feed voltage for arrays, etc.) and the frequency loop. The parameter loop is important for antenna arrays (Chapter 6). The frequency loop is important for broadband antennas (Chapter 7) and for ultrawideband or pulse antennas (Chapter 8). For large antenna structures (when the size of the impedance matrix exceeds 1000 × 1000) these loop may be rather slow. Matlab offers a wide choice of optimization routines ranging from the Optimization Toolbox to Neural Network Toolbox and a third-party Genetic Toolbox. However, none of them are employed in the present text.

8   INTRODUCTION

Table 1.2. Execution Times (Averaged): 3101 × 3101 Impedance Matrix

Machine Configuration	Operating System	Matlab	T1, min	T2, min
Pentium IV 1.7 GHz WT70-EC System Board 1 Gbyte RAM	Windows ME	Matlab 6 Release 12	4.5	2.5
Same as above	Same as above	Matlab compiler (version 2.1)	4.1	2.5
Same as above	Linux (Kernel version 2.2.19)	Matlab 6 Release 12	3.5	2.5
IBM Netfinity 6000R Series 350 4 × 700 MHz Intel Xeon CPUs 2 MB cache each	Linux (Kernel version 2.4.14) with support for "symmetric multiprocessors"	Matlab 6 Release 12	1.5	0.7

1.8. SPEED AND MAXIMUM SIZE OF THE IMPEDANCE MATRIX

Table 1.2 outlines processor time necessary to fill the impedance matrix Z and solve the system of MoM equations for the antenna shown in Fig. 1.1 (air-filled patch antenna). The structure has 3101 RWG edge elements and the 3101 × 3101 impedance matrix. The time required to fill the impedance matrix (T1) and the time required to solve the MoM equations (T2) are reported separately. For the solution of MoM equations, the built-in Matlab solver (Gaussian elimination) is used. In the last case in Table 1.2, shell scripting was used to run four Matlab processes simultaneously on a four-processor machine, thus cutting the effective time required for one process by a quarter.

The use of Matlab compiler is discussed in Chapter 2. The discussion on the maximum size of the impedance matrix is given at the end of Chapter 6. In short, although complex impedance matrixes as large as 5000 × 5000 can still be created and saved on the hard drive, it is not possible to obtain a direct solution of the matrix equation (Gaussian elimination). An `Out of memory` warning message appears that cannot be handled using the standard Matlab help recommendations. The realistic maximum size of the impedance matrix should therefore not exceed 4000 × 4000.

1.9. OUTLINE OF CHAPTERS

Chapter 2 introduces the scattering algorithm for an antenna. The goal of this chapter is to find current distribution on the antenna surface if the incident electromagnetic signal is a plane wave of given direction and polarization. The scattering of a metal plate, dipole antenna, bowtie antenna, and a slot antenna is considered. The resulting surface current distribution is the major parameter of interest. All antenna structures in this chapter are created using the mesh generator of the Matlab PDE toolbox.

Chapter 3 discusses near- and far-field parameters of an antenna. The goal of this chapter is to find the radiated field if the surface current distribution is given. Antenna directivity, gain, and effective aperture are introduced. Matlab scripts are discussed that output two- and three-dimensional radiation patterns.

Chapter 4 introduces the antenna radiation algorithm, which is the combination of the codes of Chapters 2 and 3, respectively. The goal of this algorithm is to find the antenna impedance and the radiated field if the voltage signal in the antenna feed is given. The concept of the voltage feed (delta gap) is discussed in some detail. The dipole and monopole antennas are the subjects of investigation. The corresponding antenna structures are created either using short custom Matlab scripts (mesh generators) or with the help of the Matlab PDE toolbox.

Chapter 4 also introduces the concept of the receiving antenna and shows how to calculate the received voltage when the voltage in the feed of the transmitting antenna is given. We compare the Friis transmission formula with the exact simulation and demonstrate limitations of the Friis formula with regard to loss of phase information.

Chapter 5 applies the antenna radiation algorithm to loop antennas. Single-loop antennas include electrically small-, large-, and intermediate-size loops. Center-fed helical antennas are considered, both in the normal and axial mode of operation. The antenna structures are created using the corresponding mesh generators. The generator scripts can be modified to create a tapered helix, a monopole helix on a ground plane, a cavity-backed helix, and so on.

Chapter 6 discusses the antenna arrays. We consider arrays of dipoles, monopoles, and bowties. Important concepts for an array are the multiple voltage feeds and the terminal array impedance. Two basic array types—broadside and end-fire—are introduced and investigated for a number of examples. Chapter 6 introduces the Matlab loop versus a parameter. That parameter is either array spacing for a broadside array or the incremental phase shift for an end-fire array. We show how to optimize linear end-fire arrays of dipoles using this loop.

The array meshes in Chapter 6 are mostly created using a "cloning" algorithm for the single-array element. The cloning procedure in Matlab is done very simply and requires only a few lines of the code.

Chapter 7 introduces the frequency loop and shows how to calculate antenna parameters over a frequency band. We discuss the concept of a resonant and broadband antenna, and define the antenna bandwidth. The broadband antennas studied in Chapter 7 are the bowtie antenna and the (plane) spiral antenna. The antenna structures are created using the corresponding mesh generators.

Chapter 7 describes a multiple-band antenna—the Sierpinski fractal. The antenna structure at various stages of fractal growth is created using the corresponding mesh generator. The fractal mesh requires certain modification at the edges.

Chapter 8 shows how to compute parameters of a time domain or pulse antenna using the frequency loop from Chapter 7 (frequency domain method). As an example, we consider a slot antenna of Time Domain, Co., intended for UWB pulse transmission. The first frequency loop finds the antenna radiation spectrum or, which is the same, antenna-to-free-space transfer function. The second frequency loop calculated the received voltage spectrum, or antenna-to-antenna transfer function. Based on the transfer function, the received voltage pulse is computed for two antennas separated 1 m apart and for different transmitted pulses. The algorithm is based on direct and inverse DFT.

Chapter 9 covers antenna loading and is rather short. We show that RWG edge elements are appropriate to model a resistive (or capacitive or inductive) lumped antenna load. The example considered is a loaded dipole.

Chapter 10 introduces and studies patch antennas. We consider the probe-fed patch antennas. The full-wave solution is given for air-filled patch antennas, without the dielectric. The dielectric is introduced in the approximation of an electrically thin substrate. This approximation is valid either at low frequencies or for low-epsilon dielectrics. The inviting features of this chapter are simple mesh generators for the patches on finite ground planes. Multiple and parasitic patches and patch arrays on finite ground planes can be generated and tested very quickly using Matlab mouse input.

Every chapter is accompanied by a set of problems. Asterisk after the problem number indicates higher complexity.

REFERENCES

1. C. A. Balanis. *Antenna Theory: Analysis and Design*, (2nd ed.) Wiley, New York, 1997.
2. S. M. Rao, D. R. Wilton, and A. W. Glisson. Electromagnetic scattering by surfaces of arbitrary shape. *IEEE Trans. Antennas and Propagation*, 30 (3): 409–418, 1982.
3. A. F. Peterson, S. L. Ray, and R. Mittra. *Computational Methods for Electromagnetics*. IEEE Press, Piscataway, NJ, 1998.
4. B. M. Kolundžija, J. S. Ognjanović, and T. K. Sarkar. *WIPL-D: Electromagnetic Modeling of Composite Metallic and Dielectric Structures*. Artech House, Norwood, MA, 2000.
5. J. Shin. *Modeling of arbitrary composite objects with applications to dielectric resonator antennas*. Ph.D. dissertation, The University of Mississippi, MS 38677, August 2001.
6. S. Makarov. MoM antenna simulations with Matlab: RWG basis functions. *IEEE Antennas and Propagation Magazine*, 43 (5): 100–107, 2001.
7. J. Shin, A. W. Glisson, and A. A. Kishk. Modeling of general surface junctions of composite objects in an SIE/MoM formulation. In *16th Annual Review of Progress in Applied Computational Electromagnetics*, ACES, Monterey, CA, 2000, pp. 683–690.
8. W. A. Johnson, D. R. Wilton, and R. M. Sharpe. Modeling scattering from and radiation by arbitrarily shaped objects with the electric field integral equation triangular surface patch code. *Electromagnetics*, 10 (1–2): 41–63, 1990.

2

RECEIVING ANTENNA: THE SCATTERING ALGORITHM

2.1. Introduction
2.2. Code Sequence
2.3. Creating the Antenna's Structure
2.4. RWG Edge Elements
2.5. Impedance Matrix
2.6. Moment Equations and Surface Currents
2.7. Visualization of Surface Currents
2.8. Induced Electric Current of a Dipole Antenna
2.9. Induced Electric Current of a Bowtie Antenna
2.10. Induced Electric Current of a Slot Antenna
2.11. Using the Matlab Compiler
2.12. Using Matlab for Linux
2.13. Conclusions
References
Problems

2.1. INTRODUCTION

A receiving antenna may be viewed as any metal object that scatters an incident electromagnetic field (signal). As a result of scattering, an electric current appears on the antenna's surface (Fig. 2.1a). That current in turn creates a corresponding electric field. If we cut a narrow gap in the body of an antenna, as shown in Fig. 2.1b, a voltage difference will appear across the gap. The voltage difference across the gap, or the *gap voltage*, constitutes the received signal.

12 RECEIVING ANTENNA: THE SCATTERING ALGORITHM

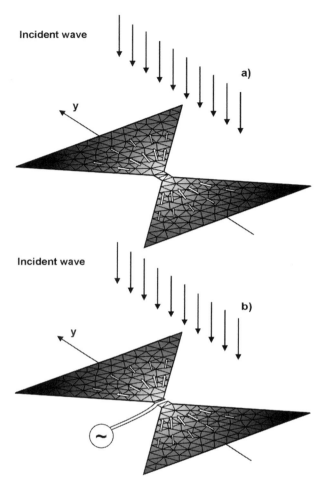

Figure 2.1. (a) Schematic of the receiving antenna (bowtie antenna); the surface current density is shown by white arrows. (b) Antenna cut to create a voltage gap.

From the viewpoint of energy transfer, an antenna in the receiving mode captures (collects) electromagnetic wave energy over a certain area. Then it extracts captured power in the voltage gap. Indeed, a very considerable amount of energy gets reflected back to free space.

The surface current distribution over the antenna surface is the most critical antenna parameter. In this chapter we will calculate surface current distributions for various antenna shapes. An incident electromagnetic signal is a plane wave whose electric field is 1 V/m. To view those shapes, make a subdirectory mesh of the Matlab directory of Chapter 2 (your current Matlab directory) and type in the Matlab command window

```
viewer platecoarse
viewer platefine
viewer dipole
viewer bowtie
viewer slot
```

Use mouse-based rotation or type `view(azim,elev)` to set the best observation angle. Use `zoom` option to see fine details of the mesh.

A classical problem—scattering from a thin metal plate at normal incidence—is considered first. Such a plate is usually employed as a test example for many numerical/experimental procedures [1–3]. Then we will test three basic antenna shapes: dipole, bowtie, and slot antenna. The presence of a narrow voltage gap does not significantly alter the scattering results.

The scattering algorithm discussed in this chapter is based on the use of the electric field integral equation (EFIE). For closed surfaces (sphere, cylinder, etc.) the stand-alone EFIE cannot be used at the resonant frequencies of the inner cavity. Instead, a combined field integral equation (CFIE) or another antiresonant technique should be employed [4].

2.2. CODE SEQUENCE

The source code in the Matlab directory of chapter 2 includes relatively short Matlab scripts `rwg1.m - rwg5.m`. These scripts reflect successive numerical steps of the moment method. Before reading this chapter, you may want to run these scripts first. The final script will display the surface current distribution for a square metal plate. Figure 2.2 shows the flowchart of the code sequence. The operation of each script is explained in some detail later in this chapter.

The code sequence in Fig. 2.2 is applicable to different antenna geometries. To replace one metal antenna object by another, it is necessary to replace the corresponding mesh file name in the starting script `rwg1.m`. Frequency, electric permittivity, and magnetic permeability are specified in the script `rwg3.m`. Interestingly, nearly the same sequence of operations is not only valid for antenna scattering but also for antenna radiation (Chapter 4, this book). The only important difference is that the antenna excitation is given by a voltage feed and not by the incident electromagnetic wave.

2.3. CREATING THE ANTENNA'S STRUCTURE

An important problem is how to create the antenna's structure. Matlab provides several ways of doing so. One is to use the built-in mesh generator of the Matlab PDE toolbox. A trial version of the PDE toolbox can be down-

14 RECEIVING ANTENNA: THE SCATTERING ALGORITHM

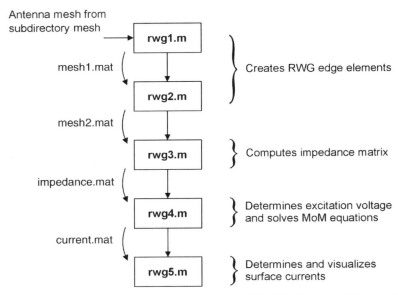

Figure 2.2. Flowchart of the scattering algorithm of Matlab's directory of Chapter 2.

loaded from the Mathworks, Inc. Web site. This mesh generator creates planar structures of any (intercepting) rectangles, polygons, and circles, using the convenient graphical user interface (GUI). Delaunay triangulation algorithm is used, along with adaptive triangle subdivision. To extend the design to a 3D structure it is usually enough to write a short Matlab script involving the z-coordinate dependency.

Another way is to identify the boundary of the antenna structure analytically. For example, the dipole can be modeled by a thin strip with four edges. Then Delaunay triangulation is applied to that structure, using Matlab function `delaunay`. To approach 3D structures, function `delaunay3` (3D Delaunay tessellation) may be used. The advantage of this approach is that we do not need the PDE toolbox and can create arbitrary 3D antenna meshes. Many examples of these operations are considered in the following text. Also otherwise created meshes can be imported to Matlab in ASCII format.

We start with the investigation of the first example—a square metal plate of infinitely small thickness [1–3]. Although such a plate is not yet the best antenna design, it is a convenient (and fast) test for many numerical/experimental procedures.

In this chapter the Matlab PDE Toolbox will be employed to create antenna meshes. If your machine doesn't have the PDE toolbox installed, you can simply use function `viewer filename` to examine structure meshes saved in subdirectory `mesh` of this chapter. Make sure that your Matlab current directory is the subdirectory `mesh`.

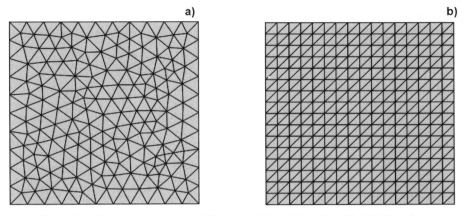

Figure 2.3. Two plate meshes: (a) Nearly equilateral triangles; (b) right triangles.

First, open the PDE toolbox (command pdetool), and draw a rectangle 1 × 1 m. An initial mesh can be generated by clicking on the button Δ or by selecting Initialize Mesh from the Mesh menu. The result is shown in Fig. 2.3a (the so-called *unstructured* mesh). In order to generate a *structured* mesh of right triangles, go to Parameters from the Mesh menu and type the maximum edge size as inf. Refine the mesh several times to obtain the result of Fig. 2.3b.

To export the mesh to the main workspace, you can select the Export Mesh option from the Mesh menu and press the ok button. The Matlab workspace will then have three arrays:

$p(2,P)$—array of Cartesian node coordinates x and y; the number of nodes is P;

$t(4,N)$—array of node numbers for each triangle;[1] the number of triangles is N;

$e(7,Q)$—array of boundary edges; the number of boundary edges is Q.

The array of boundary edges, e, is not actually needed, since it will be created automatically when building the Rao-Wilton-Glisson basis functions (script rwg1.m). The array p should be converted to a more general 3D form by adding a coordinate z as follows:

```
p(3,:)=0;
```

---

[1] The fourth row in the triangle array is the domain number. It can be ignored for a single-domain structure (e.g., a plate).

Then arrays *p* and *t* are saved in a binary *.mat file using a command

```
save filename p t
```

This binary file (antenna structure file) is further used as an input to the edge generator rwg1.m and to the rest of the code. The subdirectory mesh of the Matlab directory of Chapter 2 has two such examples for the plate: platecoarse.mat and platefine.mat with 128 and 512 triangles, respectively.

The antenna meshes are observed using function viewer('filename') or viewer filename—for example, viewer platecoarse from the same subdirectory mesh. This function enables 3D mouse-based mesh rotation (Matlab command rotate3d on).

2.4. RWG EDGE ELEMENTS

The impedance matrix discussed here was created for the Rao-Wilton-Glisson (RWG) edge elements (see Chapter 1 and [1]) and not for the triangle patches shown in Fig. 2.3. An edge element includes two triangles sharing a common edge. One of them is labeled by a plus sign and the other by a minus sign. Figure 2.4 shows three edge elements that contain the same triangle T.

There are more edges than triangles for a given structure. Therefore the number of RWG elements, M, is larger than the number of triangles, N, (usually by a factor of 1.2 to 1.4). Prior to calculation of the impedance matrix, the edge elements shown in Fig. 2.4 should be created. This is done by running the Matlab script rwg1.m. More specifically, we should identify all nonboundary edges of the antenna mesh and assign two neighbor triangles to each edge.

The discussion why we cannot use single triangles instead of the edge elements for the electric field integral equation was pursued in [1]. For the magnetic field integral equation [4], however, simple triangles (low-order boundary elements) are perfectly adequate. Unfortunately, in contrast to the electric field integral equation, the magnetic field integral equation cannot be applied to nonclosed surfaces like the plate [1,4].

The starting point for the antenna analysis is the script rwg1.m. This script reads the corresponding mesh file from the subdirectory mesh. The mesh file name should be specified at the beginning of the code by altering command load('filename'). The script counts all nonboundary (interior) edges of the mesh. For each nonboundary edge m, two triangles $T_m^{\pm}$ attached to it are found using a sweep over all triangles. The order of a triangle's numbering (plus or minus) is not important. The output of the script (saved in binary file mesh1.mat) includes the following arrays:

RWG EDGE ELEMENTS   17

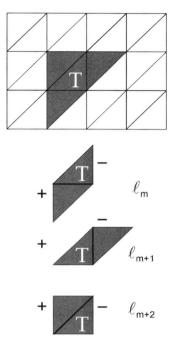

Figure 2.4. Three RWG edge elements m, m + 1, and m + 2 (with two triangles each), all containing the same common triangle T.

Edge first node number	Edge_(1,m)
Edge second node number	Edge_(2,m)
Plus triangle number	TrianglePlus(m)
Minus triangle number	TriangleMinus(m)
Edge length	EdgeLength(m)

The length of these arrays is EdgesTotal. The triangle area is calculated separately for every single patch using the vector cross-product of two side vectors. Simple space averaging of triangle's vertex points yields the triangle's midpoint (center):

Triangle area	Area(m)
Triangle center	Center(1:3,m)

The length of these arrays is TrianglesTotal.

The arrays listed above are, in principle, sufficient to build the impedance matrix of the moment method described in Chapter 1. However, the corresponding integrals can be approximated at triangle midpoints only. This is not accurate for near-diagonal terms of the impedance matrix. Moreover such

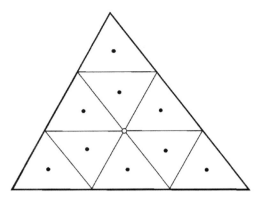

Figure 2.5. Barycentric subdivision of the primary triangle. The triangle's midpoint is shown by a white circle.

a simplified approach fails for diagonal elements, where infinite impedance values will be predicted. A number of methods for the calculation of the impedance matrix exist [1,6–9], starting with the method of the classic paper [1]. These methods employ different ways of integrating over surface triangles. Analytical approaches (line integrals and potential integrals) are accurate and fast but require extensive preliminary mathematical work.

An alternative is to use a numerical integration over a triangle [9]. If the quadrature points do not coincide with the triangle's midpoint, no separate calculations for the diagonal elements of the impedance matrix are necessary. All elements of the impedance matrix can be calculated straightforwardly, using the same formula.

Figure 2.5 shows the so-called barycentric subdivision of an arbitrary triangle [10]. Any primary triangle from Fig. 2.3 can be divided into 9 equal small subtriangles by the use of the "one-third" rule. Further we assume that the integrand is constant within each small triangle. Then the integral of a function g over the primary triangle T_m is equal to

$$\int_{T_m} g(\mathbf{r})dS = \frac{A_m}{9} \sum_{k=1}^{9} g(\mathbf{r}_k^c) \qquad (2.1)$$

where points $\mathbf{r}_k^c$, $k = 1, \ldots, 9$ are the midpoints of nine subtriangles shown in Fig. 2.5 by black circles. A_m is the area of the primary triangle. The script rwg2.m outputs the subtriangle's midpoints for each triangular patch in the following format:

  9 sub-triangle midpoints Center_(1:3,1:9,m)

The array length is again `TrianglesTotal`. After evaluation of the scripts `rwg1.m` and `rwg2.m`, a binary file `mesh2.mat` will be created, which contains all arrays discussed in the present section. In other words, it contains the necessary geometric data for the calculation of the impedance matrix.

2.5. IMPEDANCE MATRIX

The square impedance matrix determines electromagnetic interaction between different edge elements. If edge elements m and n are treated as small but finite electric dipoles, the matrix element Z_{mn} describes the contribution of dipole n (through the radiated field) to the electric current of dipole m, and vice versa. The size of the impedance matrix is equal to the number of edge elements.

The vast majority of computational work is connected to impedance matrix calculations of an antenna structure. Simultaneously the impedance matrix is the most common source of program bugs. The impedance matrix does not depend on the problem statement (radiation or scattering); instead, it depends on frequency. Frequency, as well as electric permittivity (dielectric constant) of free space ε and magnetic permeability μ, are specified in the script `rwg3.m`. Once the impedance matrix (function `impmet.m`) is programmed correctly, the rest of the work usually does not constitute any difficulties. The impedance matrix of the electric field integral equation has a typical diagonal dominance (Fig. 2.6) and is almost entirely imaginary.

The impedance matrix calculation is implemented in the form of a special function `impmet` saved in the Matlab file `impmet.m`. This function is called from the script `rwg3.m`. To speed up the calculations, the impedance function is vectorized. The vectorization process requires a few contemporary arrays calculated in the script `rwg3.m`.

Below we reproduce the derivation of the impedance matrix [1] for the RWG edge elements. Quantitatively the impedance matrix of the electric field integral equation is given by

$$Z_{mn} = l_m[j\omega(\mathbf{A}^+_{mn} \cdot \boldsymbol{\rho}^{c+}_m/2 + \mathbf{A}^-_{mn} \cdot \boldsymbol{\rho}^{c-}_m/2) + \Phi^-_{mn} - \Phi^+_{mn}] \quad (2.2)$$

where indexes m and n correspond to two edge elements; ($\cdot$) denotes the dot product. l_m is the edge length of element m; $\boldsymbol{\rho}^{c\pm}_m$ are vectors between the free vertex point, $\boldsymbol{v}^{\pm}_m$, and the centroid point, $\mathbf{r}^{c\pm}_m$, of the two triangles $T^{\pm}_m$ of the edge element m, respectively. $\boldsymbol{\rho}^{c+}_m$ is directed away from the vertex of triangle T^+_m, whereas $\boldsymbol{\rho}^{c-}_m$ is directed toward the vertex of triangle T^-_m; see Fig. 1.2 of the preceding chapter.

The vectors $\boldsymbol{\rho}^{c\pm}_m$ are not yet included in the set of key arrays discussed in the previous section. However, they are easily expressed through the known quantities using two simple formulas:

20 RECEIVING ANTENNA: THE SCATTERING ALGORITHM

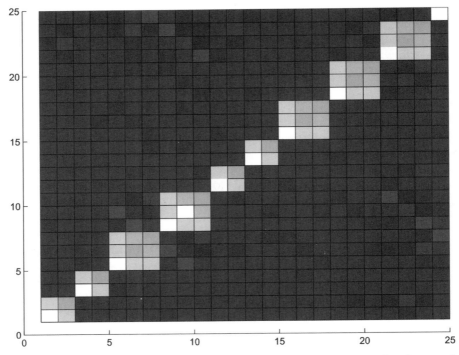

Figure 2.6. Surface plot of a block (24 × 24 elements) of an impedance matrix (imaginary part) for a plate. The white color corresponds to larger impedance magnitudes.

$$\boldsymbol{\rho}_m^{c+} = \mathbf{r}_m^{c+} - \boldsymbol{\upsilon}_m^+, \quad \boldsymbol{\rho}_m^{c-} = -\mathbf{r}_m^{c-} + \boldsymbol{\upsilon}_m^- \tag{2.3}$$

The expressions for vector $\mathbf{A}$ and scalar Φ have the form [1] ($\mathbf{A}$ is the magnetic vector potential and Φ is the scalar potential)

$$\mathbf{A}_{mn}^\pm = \frac{\mu}{4\pi}\left[\frac{l_n}{2A_n^+}\int_{T_n^+}\boldsymbol{\rho}_n^+(\mathbf{r}')g_m^\pm(\mathbf{r}')dS' + \frac{l_n}{2A_n^-}\int_{T_n^-}\boldsymbol{\rho}_n^-(\mathbf{r}')g_m^\pm(\mathbf{r}')dS'\right] \tag{2.4a}$$

$$\Phi_{mn}^\pm = -\frac{1}{4\pi j\omega\varepsilon}\left[\frac{l_n}{A_n^+}\int_{T_n^+}g_m^\pm(\mathbf{r}')dS' - \frac{l_n}{A_n^-}\int_{T_n^-}g_m^\pm(\mathbf{r}')dS'\right] \tag{2.4b}$$

where

$$g_m^\pm(\mathbf{r}') = \frac{e^{-jk|\mathbf{r}_m^{c\pm}-\mathbf{r}'|}}{|\mathbf{r}_m^{c\pm}-\mathbf{r}'|} \tag{2.4c}$$

The arrays in Eqs. (2.2) to (2.4) are (the index can be either m or n)

l_m EdgeLength(m)
$\rho_m^{c\pm}$ RHO_Plus/Minus(1:3,m)
$\rho_n^{\pm}$ RHO__Plus/Minus(1:3,1:9,m)

The array length is EdgesTotal. These arrays are also calculated in the script rwg2.m and saved in the binary file mesh2.mat.

Each of the integrals in (2.4) is calculated using the quadrature formula (2.1), which employs subtriangle arrays from the previous section. The resulting Matlab code in the script impmet.m follows Eqs. (2.1) to (2.4), and it is rather compact, since we are using symbolic vector-matrix multiplication. The script rwg3.m outputs the frequency and the impedance matrix into the binary file impedance.mat. It is worth nothing that because large meshes consume large memory, the format single can be used for the impedance matrix storage.

The script rwg3.m might require substantial CPU time (about 17 seconds to fill a 736 × 736 impedance matrix on a Pentium IV 1.7 GHz processor). Interested readers could try to modify it in a number of ways in order to speed up the impedance matrix calculation. The easiest way is to keep Eq. (2.1) for "crossing" terms only. These terms include the diagonal terms of the impedance matrix, Z_{mm}, as well as the neighboring edge elements with the common triangles. For the other edge elements, integral (2.1) may be replaced by a one-point interpolation

$$\int_{T_m} g(\mathbf{r})dS = A_m g(\mathbf{r}_m^c) \tag{2.5}$$

This modification can speed up calculations by two to four times the usual rate, but presumably accuracy is compromised.

2.6. MOMENT EQUATIONS AND SURFACE CURRENTS

The surface current density on a surface S of the plate or on other perfectly electrically conducting (PEC) structures is given by an expansion into RWG basis functions (see Eq. (1.1) of Chapter 1) over M edge elements [1]:

$$\mathbf{J} = \sum_{m=1}^{M} I_m \mathbf{f}_m, \quad \mathbf{f}_m = \begin{cases} (l_m/2A_m^+)\rho_m^+(\mathbf{r}), & \mathbf{r} \text{ in } T_m^+ \\ (l_m/2A_m^-)\rho_m^-(\mathbf{r}), & \mathbf{r} \text{ in } T_m^- \\ 0, & \text{otherwise} \end{cases} \tag{2.6}$$

If S is open, we regard $\mathbf{J}$ as the vector sum of surface currents on opposite sides of S. The units of $\mathbf{J}$ are A/m. The expansion coefficients I_m form vector

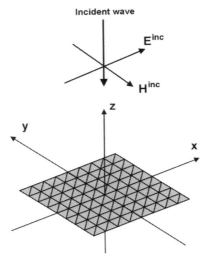

Figure 2.7. Incident field geometry for the plate.

I, which is the unique solution of the impedance equation (or the moment equation)

$$Z \cdot \mathbf{I} = \mathbf{V} \tag{2.7}$$

where the M × M impedance matrix Z is computed by running the script rwg3.m as explained in the previous section. **V** is a voltage excitation vector. When a scattering problem is considered, the voltage vector is expressed by [1]

$$V_m = l_m(\mathbf{E}_m^+ \cdot \boldsymbol{\rho}_m^{c+}/2 + \mathbf{E}_m^- \cdot \boldsymbol{\rho}_m^{c-}/2), \quad \mathbf{E}_m^\pm = \mathbf{E}^{inc}(\mathbf{r}_m^{c\pm}), \quad m = 1, \ldots, M \tag{2.8}$$

where $\mathbf{E}^{inc}$ is the electric field of an incident electromagnetic signal; (·) denotes the dot product. The voltage excitation vector is similar to the circuit voltage but has units of V·m.

For the plate we will assume that the incident signal is a *plane wave* directed perpendicular to the plate (normal incidence). The plane wave in Fig. 2.7 has only one *E*-component in the *x*-direction, $\mathbf{E}^{inc} = [E_x \ 0 \ 0]$, and this component is equal to $E_x = 1 \times \exp(-jkz)$ V/m, where $k = \omega/c$ is the wave number. If the plate is located at $z = 0$, then $\mathbf{E}^{inc} = [1 \ 0 \ 0]$ V/m. Vector [1 0 0] describes the *polarization* of the plane wave.

The script rwg4.m first determines the excitation voltage vector using Eq. (2.8). After that, the system of equations (2.7) is solved using one of the built-in Matlab matrix solvers. The simplest way to obtain the solution is by matrix inversion, I=inv(Z)*V. In practice, however, it is hardly necessary to form the explicit inverse of a matrix. A better way, from both an execution time and numerical accuracy standpoint, is to use the matrix division operator I=Z\V (left matrix divide). V should be a column vector; if this is not the case, the transpose V.' is used. This produces the solution via Gaussian elimination without forming the inverse. If the matrix size is larger than 4000 × 4000, using an iterative procedure like generalized minimum residual method (GMRES) [11] is recommended. Matlab function gmres is much faster than Gaussian elimination when the number of iterations is relatively small (10–50). The script rwg4.m outputs current coefficients I_m, vector **V**, and the corresponding frequency parameters. The script output is saved in the binary file current.mat.

2.7. VISUALIZATION OF SURFACE CURRENTS

The expansion coefficients I_m in Eq. (2.6) are not yet the surface current. The surface current density, $\mathbf{J}_k$, for a given triangle k is obtained in the form

$$\mathbf{J}_k = \sum_{m=1}^{M} I_m \mathbf{f}_m(\mathbf{r}), \qquad \mathbf{r} \text{ in } T_k \qquad (2.9)$$

A maximum of three edge elements contributes to triangle k. The script rwg5.m calculates and plots the resulting surface current density on the plate surface. Since this script is intended for arbitrary 3D structures, it uses fill3 plot and mouse rotation. Figure 2.8a shows the surface current distribution (the dominant x-component of the surface current density) that results if you run scripts rwg1.m – rwg5.m for the structure platecourse. If you refine the mesh (replace platecoarse by platefine in script rwg1.m) and then run the sequence again, the result should look like Fig. 2.8b.

The y-component of the surface current in Fig. 2.8 is small in comparison with the x-component, so it is not shown there. It might appear, at first sight, that the results of Fig. 2.8a and b are rather contradictory. The reason for this is that the surface current in the middle of the plate appears to be higher in Fig. 2.8a, whereas the surface current peaks are closer to two horizontal edges in Fig. 2.8b. As a matter of fact such a difference is due to a singular behavior of the current at two horizontal edges of the plate where its value tends to infinity in an ideal case.

The more boundary elements we have, the higher the currents at the boundary are. When the color bar extends from the minimum to the maximum current magnitude, as is the case in Fig. 2.8, then the current on the rest of the plate appears smaller in comparison with these high values. This is why the middle of the plate seems "darker" in Fig. 2.8b than in Fig. 2.8a.

To put some numbers on these data, we consider two plate cuts: at AA' and BB', as shown in Fig. 2.9a. Figure 2.9b shows the surface current distribution

24 RECEIVING ANTENNA: THE SCATTERING ALGORITHM

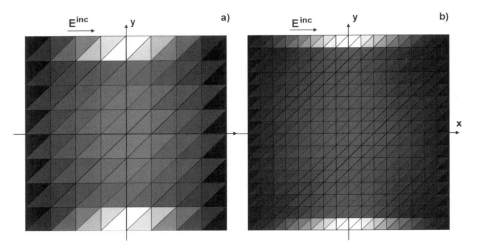

Figure 2.8. The surface current distribution, $|J_x|$, on a 1×1 m plate at a frequency of 300 MHz (plate length is equal to wavelength). The white color corresponds to higher current magnitudes. (a) 176 edge elements; (b) 736 edge elements.

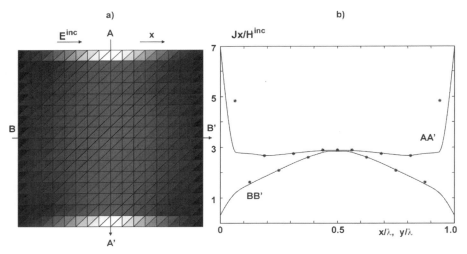

Figure 2.9. Distribution of the dominant component of surface current, $|J_x|$, across the 1λ square flat plate with 736 edge elements. (a) Two plate cuts; (b) co-polar current $|J_x|$ (solid line). The corresponding results of [2] are shown by stars.

along these cuts, normalized to the incident magnetic field (platefine.mat). For the plane wave described in Section 2.6, such a normalization implies multiplication by a factor

$$\eta = \sqrt{\mu/\varepsilon} \approx 377\,\Omega \qquad (2.10)$$

that is, by the *free-space impedance*. Also shown for comparison are the corresponding results reported in [2]. They are related to 56 edge elements instead of 736 edge elements used here.

It is evident that the present calculation and the calculation for 56 edge elements coincide well everywhere except at the plate edges. Such a discrepancy must be expected in accordance with the preceding discussion. Figure 2.9*b* is generated using the script rwg6.m included in the Matlab directory of Chapter 2. After the sequence rwg1.m - rwg5.m is complete, you may want to run this script to obtain Fig. 2.9. Do not forget to use platefine as the input antenna structure to script rwg1.m. The scripts that have a number larger than 6, such as rwg6.m and rwg7.m, are supplementary, so they are not included in the main code chart (see Fig. 2.2).

Note that a simulation for the mesh platefine.mat with 736 edge elements requires about 17 seconds to calculate the impedance matrix (script rwg3.m) and 2.5 seconds to solve the MoM equations (script rwg4.m). These results are obtained on a PC with a PIV 1.7 GHz processor and 1 Gigabyte RAM.

A way to increase the speed using the Linux version of Matlab or the Matlab compiler will be considered below in Sections 2.11 and 2.12.

2.8. INDUCED ELECTRIC CURRENT OF A DIPOLE ANTENNA

Although the case of the plate is worthy of note for the scattering purposes and radar cross-sectional modeling [5], the plate itself is hardly to be used as an antenna. A problem arises when we try to introduce the voltage gap into the plate as described in Section 2.1.

The simplest practical antenna is a thin straight wire, whose direction coincides with the direction of the incident electric field, $\mathbf{E}^{inc}$. It does not matter if we are using a cylindrical wire or a thin strip of equivalent thickness (see Chapter 4 below).

A thin long strip of 2 m length and 0.05 m width is created in the PDE toolbox. The corresponding file is dipole.m saved in subdirectory mesh. To load this file, simply type dipole in the Matlab command window. Make sure you set correctly the Matlab path. Then select the Export Mesh option from the Mesh menu and press the ok button. The rest of the commands should be

```
p(3,:)=0;
save dipole p t;
viewer dipole;
```

A binary mat file dipole.mat will then be created containing the necessary mesh data.

If your machine doesn't have the PDE toolbox, you can just type viewer dipole to examine the already created dipole mesh. Figure 2.10 shows the

26 RECEIVING ANTENNA: THE SCATTERING ALGORITHM

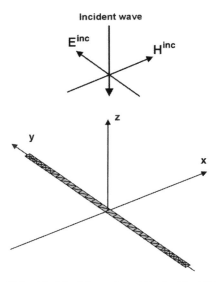

Figure 2.10. Thin-strip (dipole) mesh and the incident field geometry.

result. The plane wave incidence is shown by the arrow. Although the incidence direction coincides with that of the plate, the wave polarization in the script rwg4.m is chosen different: the E-field is now directed along the y-axis (the dipole axis).

The most popular choice is the dipole whose length (2 m in our case) is equal to half the wavelength of the incident wave. Thus, for the wavelength λ = 4 m, the frequency, f, of the incident field should be $f = c/\lambda = 75$ MHz.

Now we have to change the code sequence rwg1-rwg5.m in order to model the dipole antenna instead of the plate. To account for these changes, we return to the Matlab root directory of chapter 2. First, script rwg1.m has to be changed, where the input file name should be given in the form load('mesh/dipole'). Second, we change frequency from f=3e8 to f=75e6 in the script rwg3.m. Finally, the polarization of the incident signal must be changed. This is done in the script rwg4.m. The polarization was given by the vector Pol=[1 0 0], which corresponds to Fig. 2.7. To make it correspond to Fig. 2.10, we should type Pol=[0 1 0] in the script rwg4.m. Be sure to save the corresponding Matlab scripts after the changes are made.

The code sequence rwg1.m - rwg5.m is then executed, similar to the case of the plate. Figure 2.11a shows the resulting surface current distribution of the dipole antenna. The y-component of the current, along the dipole axis, dominates. Most noticeable is the maximum surface current in the middle of the dipole. This is exactly the place where we should introduce a voltage feed in order to convert a metal strip or wire to the real dipole antenna.

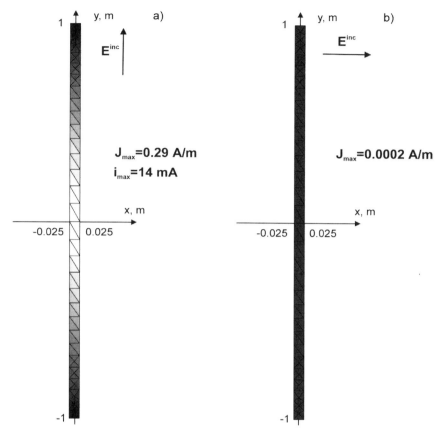

Figure 2.11. Magnitude of the surface current density along the half-wavelength strip. (a) Incident electric field is directed along the strip axis; (b) incident electric field is perpendicular to the strip axis. The white color corresponds to higher current magnitudes.

In order to estimate the magnitude of the surface current, the script rwg5.m outputs the maximum surface current density in A/m. In the case of Fig. 2.11a, the value of 0.2855 A/m is obtained. The incident electric field always has the amplitude of 1 V/m. The maximum current in the middle of the dipole is thus 0.2855 A/m × 0.05 m ≈ 14 mA.

However, if the polarization of the incident signal is changed to [1 0 0], the maximum current density magnitude drops down to 2.36×10^{-4} A/m, which is more than a factor of 1000 (Fig. 2.11b). This means that the dipole antenna is receiving nothing! Thus the dipole antenna is capable of receiving those signals whose E-field has a component parallel to the dipole axis. Such antennas are known as *polarized* antennas or antennas with *linear polarization* [12, pp. 66–69].

28 RECEIVING ANTENNA: THE SCATTERING ALGORITHM

In the script `rwg5.m`, the color bar always extends from minimum to maximum surface current magnitude. In Fig. 2.11, however, we used the same scale for the surface current density in both panels *a* and *b*. To introduce such an absolute scale the surface current should be normalized versus a given value in the script `rwg5.m`.

The dipole antenna can be scaled versus its size and frequency. For example, the 2 m long strip with the width of 5 cm at 75 MHz gives the same current distribution form as the 20 cm long strip with the width of 5 mm at 750 MHz. Indeed, the absolute current magnitudes will be different.

2.9. INDUCED ELECTRIC CURRENT OF A BOWTIE ANTENNA

The next example is a bowtie antenna shown in Fig. 2.12. The antenna length is 0.2 m and the flare angle is 90°. The width of the neck is 0.012 m. The structure is again created using PDE toolbox and `polygon tool`. The corresponding PDE file `bowtie.m` is saved in subdirectory `mesh`. To precisely adjust the polygon dimensions, one can open `bowtie.m` in the Matlab editor and change the vertex coordinates manually. Other manipulations are identical to these from the previous section.

To investigate the bowtie antenna instead of the dipole, we first set `load ('mesh/bowtie')` in script `rwg1.m`. Second, since the length of the antenna is 10 times smaller than the dipole length, we change frequency from `f=75e6` to `f=750e6` in the script `rwg3.m`. The code sequence `rwg1.m - rwg5.m`

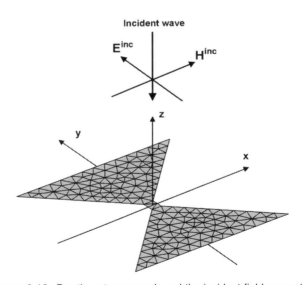

Figure 2.12. Bowtie-antenna mesh and the incident field geometry.

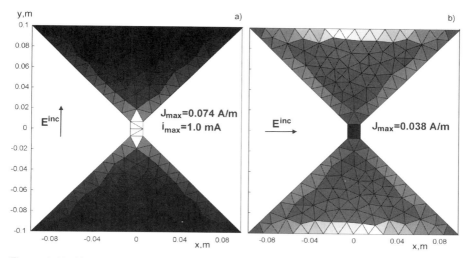

Figure 2.13. Magnitude of the surface current for the bowtie antenna at 750 MHz. (a) The incident field is directed along the antenna axis; (b) the incident field is perpendicular to the axis. The white color corresponds to higher current magnitudes.

is then executed. Figure 2.13 shows the resulting surface current distribution on the antenna surface.

If the incident signal is polarized along the y-axis, namely Pol=[0 1 0], the maximum current density appears in the area of the anticipated antenna feed (the neck area in Fig. 2.13a). However, if the incident signal is x-polarized, which is Pol=[1 0 0] in the script rwg4.m, the maximum current density appears at the edge of the antenna and is considerably smaller (Fig. 2.13b). The entire current distribution resembles that of a square plate (Fig. 2.8). The bowtie antenna is thus linearly polarized, similar to the dipole. Note that in Fig. 2.13 we use separate (maximum to minimum) scales for the surface current density in each panel (a, b).

For the bowtie antenna in Fig. 2.13a, the script rwg5.m predicts the maximum current $0.074 \times 0.012 \approx 0.012$ mA in the antenna feed. This value is considerably smaller than the corresponding result for the dipole (13 mA). The reason for the difference is that the dipole is 10 times longer than the bowtie antenna. If the dipole and the bowtie antenna will have the same lengths, comparable results must be obtained.

2.10. INDUCED ELECTRIC CURRENT OF A SLOT ANTENNA

A simplest slot antenna is a slot cut in a flat sheet metal surface shown in Fig. 2.14. The slot antenna may have a very narrow central voltage gap. To simulate this gap, we introduce a junction in the middle of the gap as shown in

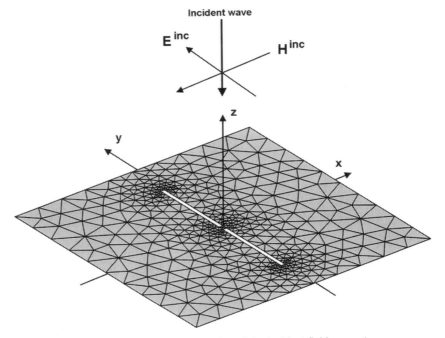

Figure 2.14. Slot-antenna mesh and the incident field geometry.

Fig. 2.14. The length of the slot in Fig. 2.14 is 2 m, the width is 0.06 m, the length of junction bridge is 0.06 m, and the plate size is 3×3 m.

To model the antenna shown in Fig. 2.14 using the Matlab PDE toolbox, a rectangle plate 3×3 m is first created. Two raw slotlike polygons are then drawn inside the rectangle (top and bottom slots) using `polygon tool`. To precisely adjust slot dimensions, we can open the corresponding m-file (`slot.m`) in the Matlab editor and change the vertex coordinates. Then we export the mesh into the main work space using the `Export Mesh` option from the `Mesh` menu. To create the binary file `slot.mat` the following commands are used again:

```
p(3,:)=0;
save slot p t;
viewer slot;
```

For the slot antenna mesh, the array of triangles, t, will provide, in its fourth column, the domain number. All triangles of the plate except two slots belong to domain 1, triangles within the top slot belong to domain 2, and triangles within the bottom slot belong to domain 3.

The domain information is contained in files `slot.m` (the PDE toolbox file) and `slot.mat` (the binary mesh file), both in subdirectory mesh. To "cut" those two slots, all triangles having a domain number greater than 1, should be omitted. Such a procedure is done by default in scripts rwg1.m and mesh/viewer.m. The output of command viewer slot is demonstrated in Fig. 2.14. The mesh has 816 triangles and 1176 edge elements, and it is the largest antenna structure considered so far.

To investigate the slot antenna, we set load('mesh/slot') in the script rwg1.m and make sure that f=75e6 in the script rwg3.m (frequency is 75 MHz). The code sequence rwg1.m - rwg5.m is again executed when the incident signal is polarized in the directions y (Pol=[0 1 0]) and x (Pol=[1 0 0]), respectively. The polarization vector must be specified in the script rw4.mat. The simulation requires about 40 seconds to calculate the impedance matrix (script rwg3.m) and about 10 seconds to solve the MoM equations (script rwg4.m). These results are for a PC with PIV 1.7 GHz processor. Figure 2.15 shows the resulting surface current distribution on the antenna surface.

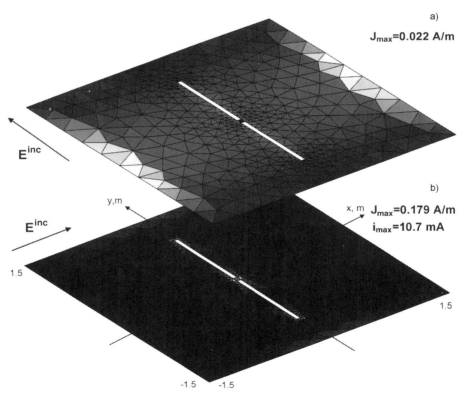

Figure 2.15. Magnitude of the surface current for the slot antenna at 75 MHz. (a) The incident field is directed along the slot axis; (b) the incident field is perpendicular to the slot. The white color corresponds to higher current magnitudes.

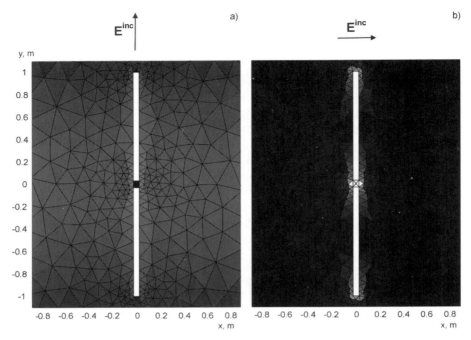

Figure 2.16. Enlarged area of the slot from Fig. 2.15. (a) The incident field is directed along the slot axis; (b) the incident field is perpendicular to the slot.

Surprisingly, nothing really happens when the incident electric field is directed along the y-axis or the slot axis (Fig. 2.15a). The current in the slot junction is very close to zero (the junction is "dark"), and the entire slot antenna just behaves like a solid plate shown in Fig. 2.8. Such a situation is exactly the opposite of the dipole excitation from Section 2.8. However, if the incident wave is x-polarized, then a large surface current flows through the gap (Fig. 2.15b). To better show these differences, Fig. 2.16a and b is provided with an enlarged area of the slot.

For the slot antenna of total length 2 m in Fig. 2.15b or in Fig. 2.16b, the script rwg5.m predicts the maximum current $0.179 \times 0.06 \approx 11$ mA in the antenna feed. This value is quite similar to the result for the dipole of the same length (14 mA). Thus the dipole and slot antennas can be considered as "electric" and "magnetic" complementary antennas: a slot can be interchanged with the dipole of the same length, if we simultaneously interchange polarization (replace the electric field by the magnetic field). The slot and the dipole therefore constitute two linearly polarized *complementary* antennas with the perpendicular polarization directions [12, pp. 618–619].

A coplanar combination of the slot and the perpendicular dipole will thus receive a signal of any polarization. Such a combination is frequently used in the patch antenna design (see Chapter 10) [13].

2.11. USING THE MATLAB COMPILER

To create C or C++ stand-alone applications from Matlab scripts of this chapter, you must have the Matlab compiler and Matlab C/C++ Math library installed on your machine. No separate C compiler is needed, since an ANSI C compiler is already included with Matlab. We will consider only one example that involves the script `rwg3.m`. This script is the most time-consuming part of the code sequence.

First, we copy scripts `rwg3.m` and `impmet.m` into the subdirectory `matlabcompiler` of chapter 2. The script code should be slightly modified in order to enable the compilation. The Matlab compiler cannot compile script files such as `rwg3.m`, nor can it compile a function m-file that calls a script [14]. Therefore we must convert the script `rwg3.m` into a function by adding a `function` line at the top of it, namely

```
function [] =rwg3
```

It also makes sense to add a line like `pause(10)` at the end of the code to prevent the DOS application from immediately closing when the program is finished. After that the code is ready for compilation. To create the stand-alone C application, the following command line should be entered [14]:

```
mcc -m rwg3
```

This command generates (1) C source codes such as `rwg3.c` plus C headers such as `rwg3.h` and (2) the executable DOS file `rwg3.exe`. You can copy the executable file into the root directory Matlab of chapter 3 and run this file using Windows Explorer. A DOS window will be open while the program is running. After the program is finished the DOS window will be automatically closed. The file `rwg3.exe` performs exactly the same task as `rwg3.m` does: it reads the structure data, calculates the impedance matrix, and saves the impedance matrix into the binary file `impedance.mat`.

It is worth to compare the execution times when using the Matlab compiler. A test with the structure `slot.mat` from subdirectory `mesh` (slot antenna with 1176 RWG elements) shows that that the processor run time for the script `rwg3.m` is 40.5 seconds (averaged over several runs). The execution time for `rwg3.exe` is 35.4 seconds (averaged over several runs). A PC with PV 1.7 GHz processor and 1 Gigabyte RAM is used for comparison. The increase in speed is therefore 13%.

Similar results were obtained for other structures, with both smaller and larger number of boundary elements. We also tested the relative performance of `rwg3.exe` and `rwg3.dll` (created with `mcc -x rwg3`) and did not find any significant difference. Moreover `rwg3.dll`, which runs in the Matlab environment, seems to be slightly slower than `rwg3.exe`. Note that the

Matlab compiler manual [14] claims the 33% increase in speed for a code that handles a smaller amount of data [14, pp. 3–3/6].

2.12. USING MATLAB FOR LINUX

The impedance code of the present chapter was tested when running the Linux version of Matlab (Kernel version 2.2.19). No changes need to be made in the code to run it under Linux. The impedance code `rwg3.m` appears to be slightly faster when it runs under Linux compared to Windows ME. Using Matlab compiler under Linux had almost no effect on the code performance.

2.13. CONCLUSIONS

In this chapter, scattering of basic antenna shapes has been investigated. We have shown how the induced electric current is distributed on the antenna surface and introduced the concept of the receiving antenna. The surface current distribution helped us to identify a most appropriate place for the antenna feed. The concept of the antenna feed will be further explored in Chapter 4.

The surface current distributions also indicate that linearly polarized antennas considered in this chapter can only receive signals of certain polarizations. In the worst case, they do not receive anything even though the electromagnetic signal is present.

The plane antenna shapes were all generated using the Matlab PDE toolbox. In a similar fashion, a variety of planar meshes can be created, including more sophisticated strip, slot, and fractal antennas. 3D surface meshes require either a custom formula for the z-coordinate, namely $z = f(x, y)$, or using 3D Delaunay tessellation implemented in Matlab function `delaunay3`.

Since 3D surface mesh capabilities of the Matlab PDE toolbox are rather limited, a number of short custom Matlab scripts (for helical antennas, spiral antennas, microstrip antennas and arrays, etc.) will be created in the following chapters.

The scattering algorithm considered in this chapter and the radiation algorithm considered in Chapter 4 constitute the background of the antenna modeling by the moment method. The scattering algorithm is used not as frequently as the radiation algorithm. Instead, a power transmission formula (or Friis transmission formula [12]) is usually employed, which gives us the magnitude of the received voltage and received power if the antenna impedance is already known (Chapter 4 below).

However, the Friis transmission formula is unable to predict the phase of the received signal. The phase information is of little importance for CW (continuous wave) transmission. However, this is important for modern ultrawideband (UWB) pulse transmission. Therefore the value of the scattering algorithm increases when we consider time domain pulse antennas. An example will be considered in Chapter 8. The scattering algorithm is also applied to study passive microwave devices such as reflectors, waveguides, and frequency selective surfaces.

It should be emphasized again that the scattering algorithm considered in this chapter and the radiation algorithm considered in Chapter 4 are almost the same. Only one Matlab script (rwg4.m) needs to be changed slightly in the basic code sequence rwg1.m-rwg5.m to account for the antenna radiation instead of scattering.

REFERENCES

1. S. M. Rao, D. R. Wilton, and A. W. Glisson. Electromagnetic scattering by surfaces of arbitrary shape. *IEEE Trans. Antennas and Propagation*, 30 (3): 409–418, 1982.
2. M. F. Catedra, J. G. Cuevas, and L. Nuno. A scheme to analyze conducting plates of resonant size using the conjugate gradient method and the fast Fourier transform. *IEEE Trans. Antennas and Propagation*, 36 (12): 1744–1752, 1988.
3. L. Gürel, K. Sertel, and I. K. Sendur. On the choice of basis functions to model surface electric current densities in computational electromagnetics. *Radio Science*, 34 (6): 1373–1387, 1999.
4. A. F. Peterson, S. L. Ray, and R. Mittra. *Computational Methods for Electromagnetics*. IEEE Press, Piscataway, NJ, 1998.
5. J. W. Crinspin and K. M. Siegel, eds. *Methods of Radar Cross-Section Analysis*. Academic Press, New York, 1968.
6. R. D. Graglia. On the numerical integration of the linear shape functions times the 3-D Green's function or its gradient on a plane triangle. *IEEE Trans. Antennas and Propagation*, 41 (10): 1448–1455, 1993.
7. C. J. Leat, N. V. Shuley, and G. F. Stickley. Triangular-patch modeling of bowtie antennas: Validation against Brown and Woodward. *IEE Proc. Microwave Antennas Propagation*, 145 (6): 465–470, 1998.
8. T. F. Eibert and V. Hansen. On the calculation of potential integrals for linear source distributions on triangular domains. *IEEE Trans. Antennas and Propagation*, 43 (12): 1499–1502, 1995.
9. J. S. Savage and A. F. Peterson. Quadrature rules for numerical integration over triangles and tetrahedra. *IEEE Antennas and Propagation Magazine*, 38 (3): 100–102, 1996.
10. Y. Kamen and L. Shirman. Triangle rendering using adaptive subdivision. *IEEE Computer Graphics and Applications* (March–April): 95–103, 1998.
11. Y. Saad. *Iterative Methods for Sparse Linear Systems*, 2nd ed. PWS Publishing, Boston, 2000.
12. C. A. Balanis. *Antenna Theory: Analysis and Design*, 2nd ed. Wiley, New York, 1997.
13. R. C. Johnson, ed. *Antenna Engineering Handbook*, 3rd ed. McGraw-Hill, New York, 1993.
14. *Matlab Compiler. User's Guide* (Version 2). MathWorks, Inc., Natick, MA, 2001.

36 RECEIVING ANTENNA: THE SCATTERING ALGORITHM

PROBLEMS

2.1. For a 1 by 1 m flat plate (structure `platefine.mat` in subdirectory `mesh`) at normal incidence of a plane wave at 600 MHz (field intensity is 1 V/m), plot the surface current distribution and compare the result to the case of plane wave incidence at 300 MHz (Fig. 2.8). The polarization of the incident wave is [1 0 0].

2.2. For a 1 by 1 m flat plate (structure `platefine.mat` in subdirectory `mesh`) at normal incidence of a plane wave at 900 MHz (field intensity is 1 V/m), plot the surface current distribution and compare the result to the case of plane wave incidence at 300 MHz (Fig. 2.8). The polarization of the incident wave is [1 0 0].

2.3. For a 2 by 0.05 m strip dipole (structure `dipole.mat` in subdirectory `mesh`) at normal incidence of a plane wave at 250 MHz, plot the surface current distribution and compare the result to the case of plane wave incidence at 75 MHz (Fig. 2.11). Estimate the maximum total current along the dipole axis. The total current is obtained as the product of the strip width and maximum surface current density. The polarization of the incident wave is [0 1 0].

2.4. Repeat Problem 2.3 if the frequency of the incident wave is 450 MHz.

2.5. Generate a mesh corresponding to a circular disk of 1 m in diameter shown in Fig. 2.17a. For a normally incident plane wave at 300 MHz, plot the surface current distribution.

2.6. Generate a mesh for a planar slot reflector shown in Fig. 2.17b and used in skeleton-slot antennas [13, pp. 17-5–17-6]. For a normally incident plane wave at 150 MHz determine the surface current distribution. Consider the cases where the polarization of the incident wave is [1 0 0] and [0 1 0], respectively.

2.7. The length of the bowtie antenna (structure `bowtie.mat` in subdirectory `mesh`) is 0.2 m. Increase the size of the antenna by a factor of 10. Save the new mesh in binary file `bowtie1.mat`, and plot the antenna structure.

2.8. The axis of a dipole antenna (structure `dipole.mat` in subdirectory `mesh`) coincides with the y-axis. Redirect the antenna so that its axis now coincides with the z-axis. Reduce the strip width to 0.025 m. Save the new mesh in the binary file `dipole1.mat` and plot the antenna structure.

2.9.* A narrowband antenna should receive and transmit maximum power at a given frequency. Create a mesh for a dipole antenna that generates maximum induced current if the frequency of incident wave is 32.5 MHz. Save the new mesh in the binary file `dipole2.mat`.

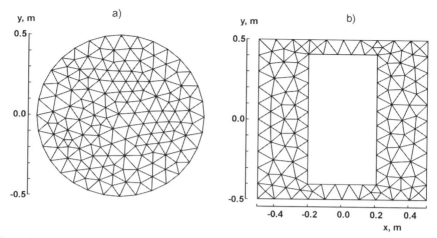

Figure 2.17. Two test planar meshes. (a) A circular disk of 1 m in diameter; (b) a slot reflector of 0.5 × 0.5 m.

Calculate the total induced current in the area of the anticipated voltage gap (antenna feed).

2.10.* A broadband antenna should receive (and transmit) the signal over a large bandwidth from 75 to 450 MHz. Possibly low variations of the induced electric current in the area of the anticipated voltage gap (antenna feed) might be required. What antenna type is more suitable for broadband applications: a 2 m long dipole or a 2 m long bowtie? Hint: Create the plot of the maximum current density versus frequency and calculate the standard deviation. Use 10 frequency points.

2.11. For a 3.0 by 3.0 m slot antenna (structure `slot` in the subdirectory `mesh`) calculate the total induced current through the slot junction when the angle between the direction of the incident wave and the x-axis is 45°. The frequency of the incident wave is 75 MHz; its polarization is [1 0 0]. Compare the obtained value to the value obtained at normal incidence in Fig. 2.15. Hint: The wave vector should be

`kv=k*[+cos(pi/4) 0 -cos(pi/4)];`

in the script `rwg4.m`.

2.12. Compare execution times of the script `rwg3.m` with and without using Matlab complier for the structure `slot.mat`.

3

ALGORITHM FOR FAR AND NEAR FIELDS

3.1. Introduction
3.2. Code Sequence
3.3. Radiation of Surface Currents
3.4. Far Field
3.5. Radiated Field at a Point
3.6. Radiation Density/Intensity Distribution
3.7. Antenna Directivity
3.8. Antenna Gain (Ideal Case)
3.9. Antenna's Effective Aperture
3.10. Conclusions
References
Problems

3.1. INTRODUCTION

A voltage in the feed of a transmitting antenna creates a surface current density that flows on the antenna surface (Fig. 3.1). This current density radiates an electromagnetic signal into the free space. The situation with the transmitting antenna is directly the inverse of the receiving antenna model considered in the preceding chapter. In Chapter 2 we calculated the induced current on the antenna surface due to an incident electromagnetic signal. In the present chapter we will calculate the radiated electromagnetic signal due to a given surface current distribution on the antenna surface.

A given surface current distribution on the surface of a metal object will be assumed in this chapter. Starting with this assumption, we will find the radi-

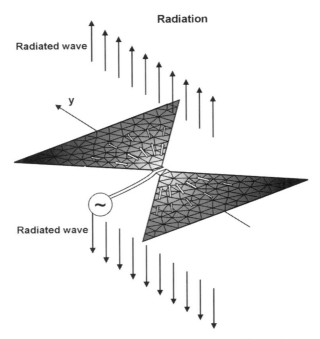

Figure 3.1. Schematic of antenna radiation (bowtie antenna). The surface current density is shown by white arrows.

ated electromagnetic signal at any spatial point, both in the far and near field of the antenna.

The feed voltage is, however, not the only source of the induced current on the antenna surface. Alternatively, the surface current may be excited by an incident (outer) electromagnetic signal, as discussed in Chapter 2. In this case we will calculate the scattered (reflected) signal as shown in Fig. 3.2. The algorithm is the same in both the cases. Only the origin of surface currents is different: either the antenna feed or an incident electromagnetic wave.

Although the present algorithm is applicable to both transmitting antennas and scattering targets, more attention will be paid to transmitting antennas. In particular, the far-field characteristics of a transmitting antenna will be introduced. A more detailed discussion of the far-field parameters (e.g., radar cross section) of the scattering problem will be pursued in the following chapters.

3.2. CODE SEQUENCE

The source code in the Matlab directory of Chapter 3 includes short Matlab scripts efield1.m - efield3.m. They calculate different parameters of the radiated field, starting with the *E*-field and *H*-field at a point (script

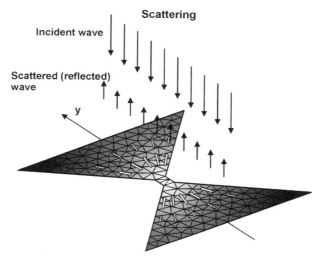

Figure 3.2. Schematic of scattered field created by induced surface currents (bowtie antenna). The surface current density is shown by white arrows.

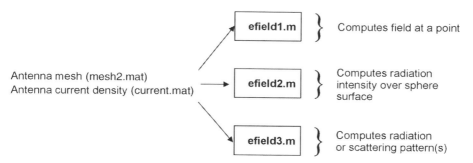

Figure 3.3. Flowchart of the Matlab scripts of chapter 3.

efield1.m) and continuing to the radiation intensity distribution over a large sphere surface (script efield2.m) and radiation patterns in the azimulthal and polar planes (script efield3.m). Figure 3.3 gives the flowchart of the code sequence.

All three scripts import two binary data files. These files are the antenna mesh with the RWG boundary elements (mesh2.mat) and the corresponding surface current distribution (current.mat). These input files are either generated by the scattering algorithm of Chapter 2, or by a radiation algorithm of Chapter 4 below. Two such input files (the dipole example from Chapter 2) are included in the Matlab directory of the present chapter.

All three scripts efield1.m - efield3.m use the function point.m that implements the dipole algorithm for the radiated field discussed below. Before reading further, you may want to run scripts efield2.m and efield3.m. These scripts will display the 3D radiation intensity distribution and a radiation pattern for a dipole antenna.

3.3. RADIATION OF SURFACE CURRENTS

Once surface currents are known on the antenna surface, a radiated electromagnetic signal in free space can be found by a number of approaches. We are generally interested in the value of the electric field, **E**, and magnetic field, **H**, at any spatial point, including both near and far field. In the near field, **E** and **H** are independent and should be calculated separately.

One method is to use the electric and magnetic field integral equations, with the observation point located somewhere outside the antenna surface [1–2]. While for the electric field integral equation only a straightforward modification of the integrals programmed in Chapter 2 is necessary (cf. Eq. (1) of Ref. [3]), the magnetic field integral equation should be programmed separately. This makes the algorithm for the radiated field rather cumbersome and slow.

Another approach, of nearly the same computational accuracy, is the so-called dipole model [4]. In the dipole model the surface current distribution for each RWG edge element containing two triangles [3] is replaced by an infinitesimal dipole, having an equivalent dipole moment or strength (Fig. 3.4). The radiated field of a small dipole is the well-known analytical expression [5, pp. 134–135; see also 6–8]. The total radiated field is then obtained as a sum of all these contributions of infinitesimal dipoles.

To find the equivalent dipole moment, we refer to notations of Ref. [3] and consider a RWG element m with two inner triangles $T_m^\pm$ adjacent to the edge of length l_m (Fig. 3.5). The dipole moment, **m**, which is the product of an effective dipole current and effective dipole length [9], is obtained by the integration of the surface current, corresponding to edge element m, over the element surface:

$$\mathbf{m} = \int_{T_m^+ + T_m^-} I_m \mathbf{f}_m(\mathbf{r}) dS = \int_{T_m^+ + T_m^-} \mathbf{f}_m(\mathbf{r}) dS = l_m I_m (\mathbf{r}_m^{c-} - \mathbf{r}_m^{c+}) \quad (3.1)$$

Figure 3.4. Schematic of the dipole approximation for a surface current distribution [4].

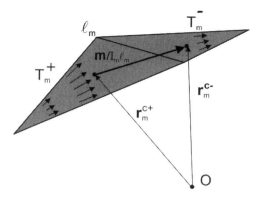

Figure 3.5. Dipole model for a surface current associated with RWG edge element m.

Here $\mathbf{f}_m(\mathbf{r})$ is the RWG basis function corresponding to element m. The surface current coefficients I_m are known from the solution of the moment equations (Chapter 2, Eq. (2.7)). The derivation of integral (3.1) is discussed in Ref. [3]. Note a misprint in Eq. (8) of Ref. [3]: it should be $\mathbf{r}_m^{c-} - \mathbf{r}_m^{c+}$ instead of $\mathbf{r}_m^{c+} - \mathbf{r}_m^{c-}$ there. The product $l_m I_m$ is associated with the dipole current, whereas the effective dipole length, l, is given by $|\mathbf{r}_m^{c-} - \mathbf{r}_m^{c+}|$ and is shown in Fig. 3.5.

The radiated magnetic and electric fields of an infinitesimal dipole located at the origin [5] are expressed at a point $\mathbf{r}$ in terms of vector notations as

$$\mathbf{H}(\mathbf{r}) = \frac{jk}{4\pi}(\mathbf{m} \times \mathbf{r})Ce^{-jkr}, \quad C = \frac{1}{r^2}\left[1 + \frac{1}{jkr}\right] \quad (3.2a)$$

$$\mathbf{E}(\mathbf{r}) = \frac{\eta}{4\pi}\left((\mathbf{M} - \mathbf{m})\left[\frac{jk}{r} + C\right] + 2\mathbf{MC}\right)e^{-jkr}, \quad \mathbf{M} = \frac{(\mathbf{r} \cdot \mathbf{m})\mathbf{r}}{r^2} \quad (3.2b)$$

Here $r = |\mathbf{r}|$; $\eta = \sqrt{\mu/\varepsilon} = 377\,\Omega$ is the free-space impedance. Equations (3.2) are the exact expressions, without any far-field approximations. Therefore they are valid at arbitrary distances from the dipole, and not only in the far field. The practical limitation of the dipole model is thus restricted by a size of the RWG edge elements. If the observation distance is on the order of dipole length (on the order of RWG element length), then the model of the infinitesimal dipole performs poorly.

The total electric and magnetic field at a point $\mathbf{r}$ are obtained as a sum

$$\mathbf{E}(\mathbf{r}) = \sum_{m=1}^{M} \mathbf{E}_m\left(\mathbf{r} - \frac{1}{2}(\mathbf{r}_m^{c+} + \mathbf{r}_m^{c-})\right) \quad \mathbf{H}(\mathbf{r}) = \sum_{m=1}^{M} \mathbf{H}_m\left(\mathbf{r} - \frac{1}{2}(\mathbf{r}_m^{c+} + \mathbf{r}_m^{c-})\right) \quad (3.3)$$

over all edge elements. Equations (3.1) to (3.3) are programmed in the Matlab function point.m, that is used in the scripts efield1.m, efield2.m, and efield3.m. Like the impedance function impmet.m, this function is vectorized to speed up the calculations.

3.4. FAR FIELD

The results could be somewhat simplified in the far field, at large radial distances r from the antenna or a scatterer. When viewed from a global perspective, the field fronts in that zone have spherical form. However, when viewed over a small range of angles, these fronts appear planar, which is one indication that they can be approximated as plane waves. **E** and **H** are perpendicular both to each other and to the direction of propagation. Namely a right-hand coordinate system is formed by the E- and H-field vectors and the direction of propagation. More precisely [5, pp. 125–126]

$$\mathbf{E}(\mathbf{r}) = \eta \mathbf{H}(\mathbf{r}) \times \frac{\mathbf{r}}{r}, \quad \mathbf{H}(\mathbf{r}) = \frac{1}{\eta}\frac{\mathbf{r}}{r} \times \mathbf{E}(\mathbf{r}) \qquad (3.4)$$

We call the field given by Eqs. (3.4) *transverse electromagnetic* TEM (to **r**) waves. These waves may be polarized in the direction perpendicular to the propagation direction but do not have a component in the direction of propagation (see [10, pp. 112–113] or [5, p. 141]).

We can give a more precise meaning of the far field (Fraunhofer region) by defining the *far-field distance* (see [5, pp. 32–34] or [10, p. 114]) in the form

$$R_f = \frac{2D^2}{\lambda} \qquad (3.5)$$

where D is the maximum dimension of the antenna and λ is the wavelength. At $r > R_f$, Eqs. (3.4) are a very good approximation.

3.5. RADIATED FIELD AT A POINT

The script efield1.m calculates radiated electric and magnetic fields at any spatial point except the antenna surface S. Prior to the calculation the input files for the antenna geometry and the surface current distribution should be specified. The Matlab directory of Chapter 3 has two such binary files mesh2.mat and current.mat. These files were calculated in Section 2.8 for a 2 m long dipole at 75 MHz. Variable ObservationPoint in the script efield1.m specifies the observation point.

Figure 3.6 shows the calculation results for two cases: when the observation point is located on the x-axis, ObservationPoint=[5;0;0] m, and when

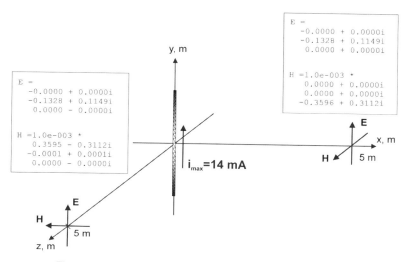

Figure 3.6. Radiated field of a dipole obtained by efield1.m.

it is on the z-axis, ObservationPoint=[0;0;5] m. Since $D = 2$ m and $\lambda = 4$ m in the present case, both points already belong to the far-field region according to Eq. (3.5).

It is seen in Fig. 3.6 that the E- and H-fields are perpendicular to each other and to the propagation direction, to a high degree of accuracy. The magnitude of the E-field is measured in V/m; the magnitude of the H-field has units of A/m. Figure 3.6 also gives a feel for the radiated signal magnitudes related to the magnitude of the total current in the dipole feed.

The cross product of **E** and **H** in Fig. 3.6 has the only component in the direction of wave propagation. Moreover the cross product of **E** and **H*** (complex conjugate) appears to be a real quantity. This product, divided by 2, is the time-average *Poynting vector* [5, pp. 35–37]. The Poynting vector at a point, **W**, gives us the average power of the radiated field, per unit area

$$\mathbf{W}(\mathbf{r}) = \frac{1}{2}\mathrm{Re}[\mathbf{E}(\mathbf{r}) \times \mathbf{H}^*(\mathbf{r})] \qquad (3.6)$$

and it has units of W/m². The calculation of the Poynting vector is done in the script efield1.m.

Note that the script efield1.m predicts exactly the same value of $|\mathbf{W}| = 41.8 \times 10^{-6}$ W/m² for two points in Fig. 3.6, that is, in both radiation directions x and z. This circumstance highlights the isotropy of the radiated field. This isotropy is essentially independent of the fact that the dipole is not a symmetric cylindrical wire but a flat thin strip in the xy-plane.

3.6. RADIATION DENSITY/INTENSITY DISTRIBUTION

The power of the radiated field per unit area as introduced by Eq. (3.6) is simultaneously the *radiation density* of the radiated signal. In the far-field region the radiation density has the only one radial component,

$$\mathbf{W}(\mathbf{r}) = W\frac{\mathbf{r}}{r} \tag{3.7}$$

which will be further denoted by W. The radiation density decreases as $1/r^2$ with increasing radius of the observation point, $\mathbf{r}$. It is therefore more convenient to introduce the *radiation intensity*, U, which is the radiation density multiplied by factor r^2, namely

$$U = r^2 W \tag{3.8}$$

The radiation intensity has units of power (W) per unit solid angle [5, p. 38] and will theoretically be the same for spheres of different radii surrounding the antenna if the sphere radius is large enough compared to the antenna size and wavelength.

The Matlab script efield2.m calculates the radiation density and/or radiation intensity over a large sphere using function point.m and Eqs. (3.6) to (3.8). The sphere mesh sphere.mat with 500 equal triangles is included into the Matlab directory of Chapter 3. The output of the script for a 2 m long dipole at 75 MHz (see Chapter 2, Section 2.8) is shown in Fig. 3.7. The sphere radius is chosen 100 m. The color bar extends from the minimum (black) to the maximum (white) intensity magnitude. Using mouse rotation, the reader can examine different angles of view. Also figure options Insert X/Y/Z-Labels can be used to identify the Cartesian axes on the plot.

The total radiated power, P_{rad} (variable TotalPower), is obtained as a sum of the products of radiation density in the middle of a sphere triangle and the triangle area. The summation is done over all 500 triangles of the sphere. In the present case the total power of approximately 8.2 milliwatts is obtained, for a 2 m long half-wavelength dipole from Matlab's Chapter 2 with the maximum current amplitude of 14 mA. The total radiated power and the antenna gain (see below) are saved in the binary file gainpower.mat.

The radiation patterns of the linearly polarized antennas (see Chapter 2 above) are often specified in terms of their *E-plane* and *H-plane* patterns (see [5, pp. 29–31]; [7, p. 585]). By definition, the *E*-plane contains the direction of maximum radiation and the electric field vector. Similarly the *H*-plane contains the direction of maximum radiation and the magnetic field vector.

The *E*- and *H*-planes are most appropriate for directional antennas whose beam patterns have unique polarizations [5, pp. 29–31]. For the dipole, however, these planes cannot be defined uniquely, since maximum radiation

ANTENNA DIRECTIVITY   47

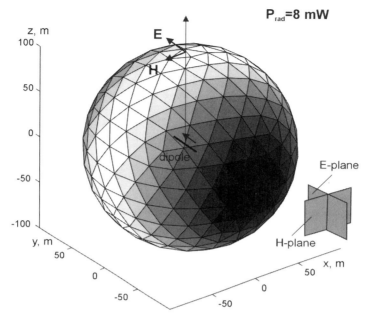

Figure 3.7. Radiation intensity distribution over a large sphere surface (output of the script `efield2.m`). The color bar extends from the minimum (black) to the maximum (white) intensity magnitude.

occurs in all directions perpendicular to the dipole axis. Figure 3.7 shows one possible arrangement if the direction of maximum radiation is chosen as the vertical direction. It is seen in Fig. 3.7 that the dipole isotropically radiates in the plane perpendicular to the dipole axis and does not radiate in the direction of the dipole axis. The same observation is valid for the receiving dipole: the dipole antenna cannot receive the signal whose propagation direction is parallel to the dipole axis.

3.7. ANTENNA DIRECTIVITY

The antenna directivity is the normalized radiation intensity calculated in dB [5, pp. 39–40, 48]:

$$D = 10 \log_{10} \frac{U}{U_0} \tag{3.9}$$

The normalization factor U_0 is the total radiated power divided by 4π (power per unit solid angle):

$$U_0 = \frac{P_{rad}}{4\pi} \tag{3.10}$$

To better understand the role of the normalization factor, let us consider an *isotropic* antenna, which radiates an equal amount of power in all directions. Then

$$\frac{U}{U_0} = 1 \tag{3.11}$$

and $D = 0\,\text{dB}$ in any direction. Any other directional antenna must therefore have a directivity above 0 dB (in the direction of maximum radiation) and below 0 dB (in the direction of minimum radiation).

A radiation pattern (or directivity pattern) is the directivity plot in terms of polar coordinates (Matlab function `polar`). Although radiation patterns are three-dimensional entities, like that shown in Fig. 3.7, they are usually measured and displayed as series of two-dimensional patterns, called cuts. The most common cuts are the elevation and azimuthal planes. Since our dipole is oriented along the *y*-axis, the *y*- and *z*-axis must be interchanged to obtain the *z*-axis orientation used in Ref. [5].

The Matlab script `efield3.m` calculates directivity patterns of the radiated field in any of the planes: *xy*, *xz*, and *yz*. This calculation is a particular case of radiation intensity calculated over the entire sphere's surface. The script `efield3.m` allows the pattern to be plotted one at a time. Alternatively, several patterns may be plotted using Matlab function `subplot`. For a dipole placed along the *y*-axis, we are interested in the *yz*-plane (elevation plane) and in the *xz*-plane (azimuthal plane), respectively. Figure 3.8 shows directivity patterns of the half-wavelength 2 m long dipole at 75 MHz (Chapter 2, Section 2.8).

A problem is usually encountered when dealing with negative directivity values (in dB). Matlab polar plot is unable to handle these negative values. A way around this problem is to add an offset of, say, +40 dB to the directivity. The polar plot may then be rescaled as it has been done in Fig. 3.8. The script `efield3.m` is using the corresponding offset (any user-defined value).

An alternative is to use a third-party (John L. Galenski III) function `polarhg`.[1] This function, also included in the Matlab directory of Chapter 3, enables negative values of directivity and many other extras. The script `efield3.m` can call this function instead of the function `polar`.

In order to calculate the directivity given by Eq. (3.9), we should know the total radiated power, P_{rad}. This requires a numerical integration of the flux of radiated energy (Poynting vector) over a large sphere. In some cases such an integration is not possible, and the total power P_{rad} remains unknown. Then the radiation pattern is normalized to its maximum value. Ref. [5]

---

[1] Dr. Paul Harms of Georgia Tech kindly sent this script to the author.

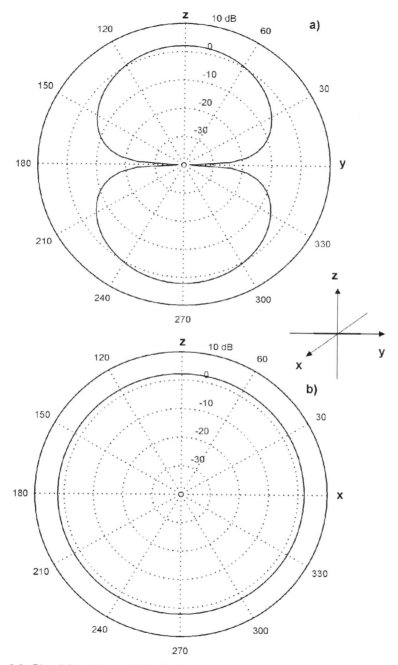

Figure 3.8. Directivity patterns of the dipole antenna in the yz- and xz-planes (output of the script efield3.m).

includes many examples of this kind. Such a normalization is obviously simpler, but it does not allow us to calculate the antenna gain considered in the next section.

Two kinds of antenna pattern shapes are given special names. The first is an *isotropic pattern*, which is the same in all directions. No antenna can have a truly isotropic pattern, but it is often a convenient idealization. An *omnidirectional pattern* is rotationally symmetric about an axis, often called the azimuthal axis.

Figures 3.7 and 3.8 show that the dipole is an omnidirectional antenna. Omnidirectional antennas are often used in broadcast applications, since they provide uniform coverage in all directions around them.

3.8. ANTENNA GAIN (IDEAL CASE)

The calculation of the antenna gain essentially completes the antenna analysis. The gain of an antenna is closely related to the directivity. For ideal antennas studied in this and following chapters, the gain is just the maximum directivity expressed in dB. The gain tells us how "directional" the specific antenna type is. Higher gains correspond to highly directional antennas (e.g., Yagi-Uda or reflector). Often the observation direction is not specified and "directivity" means the maximum directivity. In such cases directivity and gain both have equivalent meaning [5, pp. 58–60].

Since the maximum directivity can occur in any direction, especially for sophisticated antenna structures, the gain calculations are done in the script efield2.m. using the formula (loss antennas only)

$$G = 10\log_{10}\frac{\max(U)}{U_0} \qquad (3.12)$$

where the maximum is calculated over all possible directions. For the drill example of this chapter (a half-wavelength 2 m long strip at 75 MHz) the script efield2.m outputs the logarithmic gain of 2.15 dB. It might be interesting to note that the theoretically predicted gain of a half-wavelength infinitely thin dipole antenna ([5, p. 163] or [11, pp. 4–12]) is exactly (!) equal to this value.

Two values of the gain are used simultaneously: the logarithmic gain according to Eq. (3.12) and the linear gain

$$G = \frac{\max(U)}{U_0} \qquad (3.13)$$

For the dipole, the linear gain is obtained as 1.64, and the logarithmic gain is thus $10\log_{10}(1.64) = 2.15$ dB. It is worth remembering these values. Various antenna gains are usually compared to the gain of the half-wavelength dipole.

Narrowband directional antennas have much higher gain, up to 30 dB and even more. For an antenna array, the gain may be as high as 60 dB. The antenna gain generally increases with increasing frequency. For antenna elements with losses, the antenna gain includes one more important mechanism, namely ohmic losses in imperfect conductors. A more general definition of gain is given in [5, pp. 58–60] or [11, pp. 1–5].

3.9. ANTENNA'S EFFECTIVE APERTURE

The antenna's aperture is a parameter that represents the electrical surface area that an antenna presents to an incoming wave. This is rather a scattering, and not the radiation parameter. In particular, for the receiving antenna we are interested in determining the received power for a given incident plane wave field. Finding received power is important for the radio system link equation (Friis transmission formula, to be discussed in the next chapter).

We expect that the received power will be proportional to the power density, or Poynting vector of the incident wave. Since the Poynting vector has dimensions of W/m², the proportionality constant must have units of area. Thus we write

$$P_{\text{rec}} = A_e W \tag{3.14}$$

where A_e is defined as the *effective aperture area* of the receiving antenna. The effective aperture area has dimensions of m², and can be interpreted as the "capture area," intercepting part of the incident power density radiated toward the receiving antenna.

The maximum effective aperture area can be shown to be related to the (linear) gain of the antenna as [5, p. 86]

$$A_e = \frac{G\lambda^2}{4\pi} \tag{3.15}$$

where λ is wavelength. For the dipole example of this chapter, $\lambda = 4$ m, $G = 1.64$, and therefore

$$A_e = 2.1 \, \text{m}^2 \tag{3.17}$$

This value is considerably larger than the actual area of the antenna surface (a 2 m long and 5 cm wide strip has the surface area of 0.1 m²). Figure 3.9 shows schematically the actual antenna area and the effective aperture area (capture area) of the half-wavelength dipole. For directional antennas like the Yagi-Uda antenna the capture area can be even larger.

An ideologically similar parameter, widely used in radar technology, is the radar cross section [12,13]. Physically, a radar cross section is the measure of

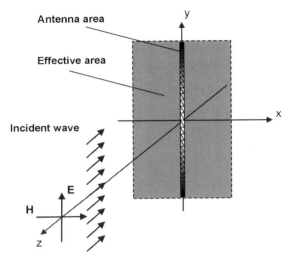

Fig. 3.9. Effective aperture area of the half-wavelength dipole.

a target's ability to reflect radar signals in the direction of the same or another radar. In other words, for the antennas, we are interested in the captured power over a certain area, whereas for the radars, we are interested in the reflected power over a certain area. Radar cross sections of various targets are considered in [5, pp. 90–98].

3.10. CONCLUSIONS

In this chapter we discussed the algorithm for the calculation of a radiated electromagnetic field due to a given surface current distribution on the surface of a metal object. The origin of the surface current does not play any role. Therefore the algorithm is equally applied to antenna radiation (surface current induced by a voltage feed) and antenna scattering (surface current induced by an incident signal).

A dipole model is employed to find the radiation from the RWG edge elements. This way we can perform fast and accurate calculations both in the near and far field. The most important antenna characteristics in the far field were introduced such as antenna directivity and gain.

The drill example of the present chapter was the half-wavelength dipole and the associated current distribution. This current distribution is identical to the current distribution of the radiating dipole considered in Chapter 4, except for the difference in magnitudes. Therefore, in studying this current distribution, we were able to quantitatively determine some parameters of the half-wavelength transmitting dipole antenna such as the gain of 2.15 dB and effective aperture area of 2.1 m^2.

REFERENCES

1. A. J. Poggio and E. K. Miller. Integral equation solutions of three-dimensional scattering problems. In R. Mittra, ed., *Computer Techniques for Electromagnetics*, 2nd ed. Pergamon Press, Oxford, 1973, pp. 159–264.
2. A. F. Peterson, S. L. Ray, and R. Mittra. *Computational Methods for Electromagnetics*. IEEE Press, Piscataway, NJ, 1998.
3. S. M. Rao, D. R. Wilton, and A. W. Glisson. Electromagnetic scattering by surfaces of arbitrary shape. *IEEE Trans. Antennas and Propagation*, 30 (3): 409–418, 1982.
4. C. J. Leat, N. V. Shuley, and G. F. Stickley. Triangular-patch modeling of bowtie antennas: Validation against Brown and Woodward. *IEE Proc. Microwave Antennas Propagation*, 145 (6): 465–470, 1998.
5. C. A. Balanis, *Antenna Theory: Analysis and Design*, 2nd ed. Wiley, New York, 1997.
6. J. D. Kraus. *Antennas*. McGraw Hill, New York, 1950.
7. K. R. Demarest. *Engineering Electromagnetics*. Prentice Hall, Upper Saddle River, NJ, 1998.
8. Akira Ishimary. *Electromagnetic Wave Propagation, Radiation, and Scattering*. Prentice Hall, Upper Saddle River, NJ, 1991.
9. J. D. Kraus. *Electromagnetics*, 4th ed. McGraw Hill, New York, 1992.
10. D. M. Pozar. *Microwave and RF Design of Wireless Systems*. Wiley, New York, 2001.
11. R. C. Johnson, ed. *Antenna Engineering Handbook*, 3rd ed. McGraw-Hill, New York, 1993.
12. E. F. Knott, J. F. Shaeffer, and M. T. Tuley. *Radar Cross Section: Its Prediction, Measurement and Reduction*. Artech House, Dedham, MA, 1985.
13. J. W. Crispin and K. M. Siegel, eds. *Methods of Radar Cross-Section Analysis*. Academic Press, New York, 1968.

PROBLEMS

3.1. For a 2 m long dipole at 75 MHz, determine the magnitude of the radiated electric field along the z-axis at the distances $z = [0.1:0.1:1]$m from the antenna. Plot the magnitude to scale as a function of distance from the antenna.

3.2. For a 2 m long dipole at 75 MHz, determine the magnitude of the radiated magnetic field along the z-axis at the distances $z = [0.1:0.1:1]$m from the antenna. Plot the magnitude to scale as a function of distance from the antenna.

3.3. Repeat Problem 3.1 if $z = [0.25:0.25:5]$m.

3.4. Repeat Problem 3.2 if $z = [0.25:0.25:5]$m.

3.5. For a 2 m long dipole a 75 MHz, find the components of the radiated electric field:

54 ALGORITHM FOR FAR AND NEAR FIELDS

    a. At a point $\mathbf{r} = [0.2 \quad 0.2 \quad 0.2]$m
    b. At a point $\mathbf{r} = [2 \quad 2 \quad 2]$m.

3.6. For a 2 m long dipole a 75 MHz, find the components of the radiated magnetic field:
    a. At a point $\mathbf{r} = [-0.2 \quad 0.1 \quad 0.1]$m
    b. At a point $\mathbf{r} = [-2 \quad 1 \quad 1]$m.

3.7.* In the far field of an antenna, the magnitude of electric and magnetic fields are related by the formula $|\mathbf{H}(\mathbf{r})| = 1/\eta |\mathbf{E}(\mathbf{r})|$, where η is the free-space impedance (approximately 377 Ω). For a 2 m long dipole at 75 MHz, check how well this formula is satisfied:
    a. At a point $\mathbf{r} = [0.5 \quad 0.5 \quad 0.5]$m
    b. At a point $\mathbf{r} = [50 \quad 50 \quad 50]$m.
    Explain the disagreement at a relatively small distance from the antenna.

3.8. For a 2 m long dipole at 75 MHz, determine the magnitude of the Poynting vector along the z-axis at the distances $z = [0.1:0.1:1]$m from the antenna. Plot the magnitude to scale as a function of distance from the antenna.

3.9. Repeat Problem 3.8 if $z = [0.25:0.25:5]$m.

3.10.* In the far field of an antenna, the Poynting vector can be found using the formula

$$\mathbf{W}(\mathbf{r}) = \frac{1}{2\eta} |\mathbf{E}(\mathbf{r})|^2 \frac{\mathbf{r}}{r} \quad \text{(plane wave approximation)}$$

For a 2 m long dipole at 75 MHz, compare the magnitude of the Poynting vector obtained using this formula and Eq. (3.6):
    a. At a point $\mathbf{r} = [0.25 \quad 0.25 \quad 0.25]$m
    b. At a point $\mathbf{r} = [25 \quad 25 \quad 25]$m.
    Explain the disagreement at a small distance from the antenna.

3.11. For a 2 m long dipole at 75 MHz and maximum current of 14 mA, determine the total radiated power crossing the surface of the sphere surrounding the dipole when:
    a. Sphere radius is 1.2 m
    b. Sphere radius is 120 m
    c. Sphere radius is 1200 m.
    Can you make any conclusion about the numerical error in power calculations?

3.12. For a 2 m long dipole at 75 MHz and maximum current of 14 mA, determine the (maximum) antenna gain (logarithmic and linear) over the surface of the sphere surrounding the dipole when:

  d. Sphere radius is 1.2 m
  e. Sphere radius is 120 m
  f. Sphere radius is 1200 m.

  Can you make any conclusion about the numerical error in gain calculations?

3.13. For a 2 m long dipole at 75 MHz, determine the total radiated power and directivity patterns in the elevation and azimuthal planes if the given surface current distribution (file `current.mat`) is multiplied by a factor of 10.

3.14.* Two parallel 2 m long identical dipoles at 75 MHz form an antenna array. The separation distance of 5 m is large enough so that the mutual coupling is negligibly small. Obtain array gain and plot the directivity patterns of the array in the elevation and azimuthal planes. Hint: Modify function `point.m` to account for the surface current distribution of two equal dipoles separated by 5 m.

4

DIPOLE AND MONOPOLE ANTENNAS: THE RADIATION ALGORITHM

4.1. Introduction
4.2. Code Sequence
4.3. Strip Model of a Wire
4.4. Feed Model
4.5. Current Distribution of the Dipole Antenna
4.6. Input Impedance
4.7. Monopole Antenna
4.8. Impedance of the Monopole
4.9. Radiation Intensity, Radiated Power, and Gain
4.10. Radiation Resistance and Delivered Electric Power
4.11. Directivity Patterns
4.12. Receiving Antenna
4.13. Friis Transmission Formula
4.14. Conclusions
References
Problems

4.1. INTRODUCTION

The task of the scattering algorithm in Chapter 2 was to find the surface current distribution on the antenna surface induced by an incident electromagnetic wave. The task of the antenna radiation algorithm discussed in this chapter is to find the surface current distribution due to an applied voltage in the antenna feed (Fig. 4.1). After the surface current distribution is

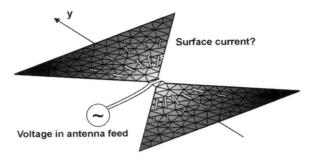

Figure 4.1. Task of the radiation algorithm.

obtained, the radiated field of an antenna is found using the algorithm of Chapter 3.

The scattering algorithm developed in Chapter 2 can easily be modified for the antenna radiation. The major challenge related to transmitting antennas is the antenna feed, programmed in the script rwg4.m. Whereas for the receiving antennas and EM studies the excitation is simply the field of an incident plane wave, in a transmitting antenna the excitation is generally a voltage source on a wire or plate conductor.

Wire antennas are traditionally studied in terms of a one-dimensional segment model (numerical code NEC [1–6]). The theory behind NEC requires a special integral equation and a special set of basis functions. There are two different models that we must keep track of: one model for conducting surfaces and another model for wires [2]. More problems arise when wires and surfaces are joined together [2]. Although NEC is rather fast, these circumstances are very inconvenient.

To avoid the development and programming of two separate algorithms, we must investigate the potential of RWG boundary elements for modeling wire antennas. A wire is represented with the use of a thin-strip model having one or two RWG edge elements per strip width. The results obtained are encouraging. Faithful reproduction of the surface current distribution, input impedance, and gain are observed for dipole and monopole antennas. Therefore both patch and wire antennas will be described here using the same numerical method—RWG edge elements [7].

4.2. CODE SEQUENCE

Successive numerical steps, necessary to calculate an antenna, are implemented in the relatively short scripts rwg1.m - rwg5.m. These scripts, except for rwg4.m, are identical to those from Chapter 2 and can be found in the Matlab directory of the present chapter. Before reading the chapter, you may

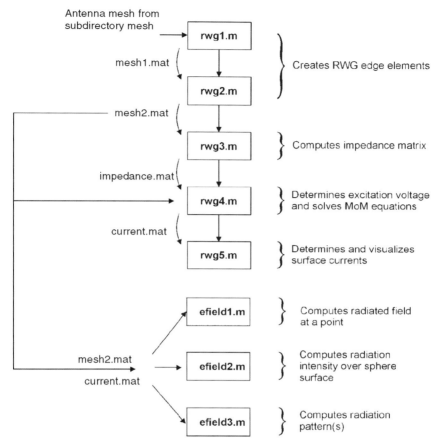

Figure 4.2. Flowchart of the complete radiation algorithm.

want to run these scripts. The final script displays the surface current distribution for a half-wavelength dipole due to a feed voltage of 1 V. The antenna mesh name should be specified in the script rwg1.m. Calculations of the input impedance and radiation resistance are done in the script rwg4.m.

After the first code sequence is complete, scripts efield1.m, efield2.m, and efield3.m provide radiated signal at a point, radiation patterns of the antenna including 3D patterns, and the antenna gain. These scripts are identical to the scripts of Chapter 3. The present code sequence (Fig. 4.2) is applicable to arbitrary metal antennas, not only to dipoles or monopoles, and will be used in the rest of the text. Figure 4.2 indicates that the complete radiation algorithm is none other than a straightforward combination of the algorithm of Chapter 2, modified by the antenna feed, and of the algorithm of Chapter 3.

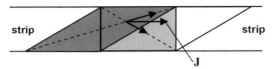

Figure 4.3. Thin-strip discretization. Two RWG elements form a current vector J directed exactly along the strip.

4.3. STRIP MODEL OF A WIRE

As far as impedance and radiation pattern are concerned, a thin cylindrical antenna with a noncircular crosssection behaves like a circular cylindrical antenna with an equivalent radius. For a thin strip, the radius of the equivalent cylindrical wire is given by ([8, pp. 4–10]; see also [1, p. 456])

$$a_{eqv} = 0.25s \qquad (4.1)$$

where s is the strip width. A typical RWG boundary element assembly of a strip is shown in Fig. 4.3. There is only one edge of the triangle across the strip. The two adjoining RWG edge elements, shown in Fig. 4.3, are able to support a uniform electric current **J** along the strip axis; this satisfies the most important assumption of the thin-wire theory [1–3].

You can use the Matlab script strip.m in subdirectory mesh to create a uniform triangular mesh for a strip of arbitrary length and arbitrary width. 2D Delaunay triangulation is employed to create the mesh. Type strip in the Matlab command window to see the result. The mesh is saved in the output binary fime strip.mat. The script strip.m is actually more flexible than the PDE toolbox, since we can control the number of boundary triangles. It provides the reader with the first example of the analytical description of the antenna structure.

The subdirectory mesh of the Matlab directory of Chapter 4 also contains a pre-generated strip mesh strip2.mat. This mesh was created using the Matlab PDE toolbox. Different strip meshes are shown in Fig. 4.4. All meshes have the same length to width ratio (100) but differ in the discretization accuracy along the strip axis. Additionally the mesh strip2.mat has finer discretization toward the ends. The strip meshes can be visualized using the function viewer filename, such as viewer strip2, from the same subdirectory mesh.

4.4. FEED MODEL

To account for a voltage source instead of an incident wave, we should introduce a feed model into the antenna structure [1,3,5]. An antenna is usually fed

FEED MODEL 61

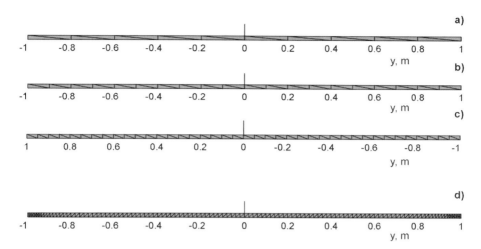

Figure 4.4. Output of `strip.m`: (a) Strip with 20 triangles, 1:100; (b) strip with 40 triangles, 1:100; (c) strip with 80 triangles, 1:100. (d) Pregenerated mesh `strip2.mat` (244 triangles, 1:100).

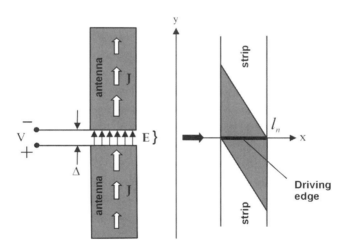

Figure 4.5. Feeding edge model. Black arrows show the electric field direction in the antenna gap. White arrows show the direction of the surface current on the antenna surface.

by a conventional transmission line through two electrically close terminals. This means that an ideal voltage generator is connected across a gap with a small width, along the antenna—see Fig. 4.5.

There are several ways to describe the gap field [1,3]. This problem has received a considerable amount of attention in the literature [1–6,9–11]. The simplest (and often less accurate) way, which is ideally suited for RWG edge elements, is the so-called delta-function generator or the feeding edge model [3,9–11]. In short, this model assumes a gap of negligible width, Δ. If the voltage across the gap is V (from positive to negative terminal), then the electric field within the gap becomes (Fig. 4.5)

$$\mathbf{E} = -\nabla\varphi = \frac{V}{\Delta}\mathbf{n}_y \qquad (4.2)$$

where φ is the electric potential. When Δ tends to zero, Eq. (4.2) predicts infinite values within the gap. That is, the delta-function approximation

$$\mathbf{E} = V\delta(y)\mathbf{n}_y \qquad (4.3)$$

Equation (4.3) simply states that the integral of the electric field over the gap is equal to the applied voltage, namely

$$\int E_y dy = V \qquad (4.4)$$

It is convenient to associate the gap with an inner edge n of the boundary element structure. There is only one RWG element corresponding to that edge (Fig. 4.5). Thus the "incident" electric field will be zero everywhere except for one RWG element, n.

Instead of Eq. (2.8) of Chapter 2, we take for the excitation voltage of the moment equation

$$V_{m=n} = \int_{T_n^+ + T_n^-} \mathbf{E} \cdot \mathbf{f}_n dS = V \int_{T_n^+ + T_n^-} \delta(y)\mathbf{n}_y \cdot \mathbf{f}_n dS = l_n V \quad \text{for edge element } m = n$$

$$V_m = \int_{T_m^+ + T_m^-} \mathbf{0} \cdot \mathbf{f}_m dS = 0 \qquad \text{otherwise} \qquad (4.5)$$

Equation (4.5) uses the fact that a component of the RWG basis function $\mathbf{f}_n$ normal to the edge is always equal to one [7]. In the following discussion, we also assume that the feeding voltage, V, is equal to 1 V. In other words, the feeding voltage is a cosine function of time having a phase of zero and an amplitude equal to 1 V.

In the command-line sequence of Chapter 2

`rwg1.m; rwg2.m; rwg3.m; rwg4.m; rwg5.m`

only the Matlab script `rwg4.m` needs to be changed to account for Eq. (4.5). Everything else is directly adopted from the scattering analysis. The Matlab script `rwg4.m` always identifies the feeding edge as being the closest one to the origin (cf. Fig. 4.4).

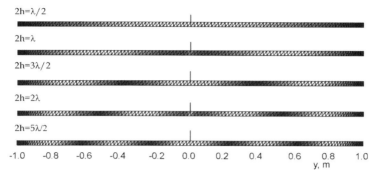

Figure 4.6. Magnitude of the surface current along the strip `strip2.mat` at different frequencies. The strip length is 2h. The white color corresponds to higher current magnitudes. The feeding edge (in the middle) is shown by a bar.

4.5. CURRENT DISTRIBUTION OF THE DIPOLE ANTENNA

The output of the command line sequence at the end of Section 4.4 is the surface current distribution along the strip. The magnitude of the surface current is given for each triangular patch. Figure 4.6 shows the output at five different frequencies for the structure `script2.mat`. The starting frequency corresponds to a half-wavelength dipole (total dipole length, 2h, is equal to a half wavelength). In our case $h = 1$ m and the frequencies are 75, 150, 225, 300, and 375 MHz, respectively. A periodic current distribution along the dipole is observed with multiple maxima and minima at higher frequencies. The periodic current distribution is typical for dipoles and similar resonant wire antennas [1, p. 156].

There are certain frequencies at which the dipole behavior becomes especially attractive. They correspond to a half-wavelength dipole and other dipoles, whose length is an odd number of half wavelengths:

$$2h = \frac{\lambda}{2}, \frac{3\lambda}{2}, \frac{5\lambda}{2}, \ldots$$

At these frequencies the dipole impedance is predominantly a resistance on the order of 50 to 200 Ω (see the next section). Therefore it is easy to design efficient matching networks that allow the dipole antennas to be driven by amplifying circuits. The half-wavelength dipole is by far the most popular antenna choice, especially in the area of wireless communications (see [12, pp. 70–75] and [13, ch. 4]).

At the same time, an infinitely thin dipole becomes a very impractical antenna at the frequencies corresponding to

$$2h = \lambda, 2\lambda, 3\lambda, \ldots$$

64　DIPOLE AND MONOPOLE ANTENNAS: THE RADIATION ALGORITHM

At these frequencies the current is almost zero at the feed point. The input impedance becomes very high and reactive. Then it is very difficult to design the corresponding matching network. However, if the dipole has a finite radius (which corresponds to a strip of a finite width), the feed current always attains finite values [1]. Figure 4.6 indicates some nonzero values of the feed current at $2h = \lambda$, $2h = 2\lambda$ seen as narrow, relatively bright spots at the feed point.

For a thin wire or a thin strip, we expect that the longitudinal component of the current in the direction of the strip dominates the transversal component. The complementary script rwg6.m allows us to obtain the corresponding quantitative estimate. A uniformly spaced grid is introduced along the strip axis (usually the y-axis). For each grid point the nearest triangle patch is found. The surface current in the middle of that triangle (see the script rwg5.m and its description in Chapter 2) is assigned to the current at the grid point. The script rwg6.m outputs both axial and transversal components of the surface current. Figure 4.7 shows the script output for the structure strip2.mat at 75 MHz (half-wavelength dipole). The maximum magnitude of the transversal current component appears approximately 200 times smaller than the magnitude of the longitudinal component. This result, also tested at various frequencies and for different structures, is critical for the subsequent analysis. It

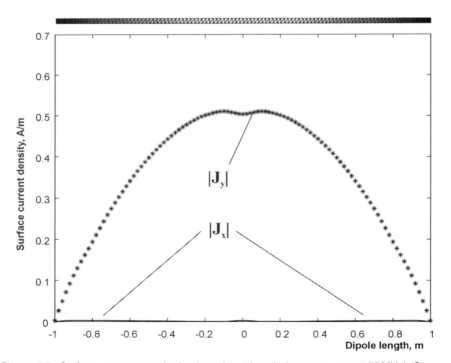

Figure 4.7. Surface current magnitude along the strip axis (strip2.mat at 75 MHz). Stars—axial component $|J_y|$; solid line—transversal component $|J_x|$.

shows that the RGW edge elements do support a directional electric current along the strip, even if we have one or two RWG elements per strip width. Therefore, instead of a thin-wire model, we can use the thin-strip model without considerable increase in CPU time. The advantage of the strip model is that a separate set of MoM equations, corresponding to the wire, becomes unnecessary.

Another positive point is the relatively simple coupling model between the strip mesh and a metal ground plane mesh. The coupling edge model will be considered later in this chapter using the example of a base-driven monopole on a finite ground plane.

4.6. INPUT IMPEDANCE

One major parameter of interest is the antenna input impedance. Once the impedance is known, other antenna parameters such as return loss can easily be obtained. The input impedance is defined as the impedance presented by an antenna at its terminals or the ratio of the voltage to current at a pair of terminals. The feeding edge model determines the impedance to be the ratio of the feeding voltage ($V = 1\,\text{V}$) to the total current normal to the feeding edge, n. In the expansion of surface currents over RWG basis functions (Chapter 2, Eq.(2.6))

$$\mathbf{J} = \sum_{m=1}^{M} I_m \mathbf{f}_m \qquad (4.6)$$

only the basis function $\mathbf{f}_{m=n}$ will contribute to the impedance calculation since no other basis functions have a component normal to the edge n. Moreover, since a component of the RWG basis function $\mathbf{f}_n$ normal to the edge is always equal to one [7], the total normal current through the edge is given by

$$l_n I_n \qquad (4.7)$$

where l_n is the edge length. The antenna impedance is simply

$$Z_A = \frac{V}{l_n I_n} = \frac{V_n}{l_n^2 I_n} \qquad (4.8)$$

according to Eq. (4.5). Input impedance is measured in ohms, and it is a complex quantity in general. Indeed, more sophisticated definitions of the impedance can always be proposed based on surface current averaging over an area around the feeding edge. However, they will not be considered in the following text.

Table 4.1. Input Impedance of Strip and Wire Models for a Half-wavelength 2 m Long Dipole at 75 MHz

Model	Input Impedance, Ω
Strip.m—20 triangles	$88 + j \times 34$
Strip.m—40 triangles	$88 + j \times 41$
Strip.m—80 triangles	$87 + j \times 44$
Strip2.mat—244 triangles	$88 + j \times 47$
Wire with 19 segments (SuperNEC)	$90 + j \times 53$
Wire with 39 segments (SuperNEC)	$92 + j \times 50$

Note: The strip width is 0.02 m. The equivalent wire radius is 0.005 m. A WIPL-D simulation gives $85 + j \times 44 \, \Omega$ (see Ref. 4 of Chapter 1).

Input impedance calculations are performed in the script rwg4.m using Eq. (4.8). The corresponding code line is inserted just after the solution of moment equations for the current expansion coefficients. Table 4.1 gives the input impedance of four different strip meshes outlined in Fig. 4.4 at a frequency of 75 MHz, assuming the center feed. That exactly corresponds to a half-wavelength dipole. For the purposes of comparison, we present similar results obtained by a NEC wire solver. They are computed using SuperNEC of Poynting Software Pty Ltd [4] and a 2 m long wire with an equivalent radius of 0.005 m. The wire is divided into 19 and 39 segments, respectively. Option "thin wire kernel" is used. Table 4.1 indicates a relatively good agreement in the input impedance. Also the reported data come in close proximity to the data measured by King [14] at similar antenna parameters.

In [10] it was brought to our attention that the edge feed must be constructed symmetrically so that "the triangle to the left of the feed point is the mirror image of the one to the right of the feed point." We have tested both symmetric and nonsymmetric triangles and did not find a significant difference. This is an indication that the feed symmetry is a desired, but probably not necessary, condition at least for thin strips or dipoles.

It is worth noting that an increase in the number of segments in SuperNEC does not guarantee convergence of the input impedance. This is in agreement with the well-known fact that the thin-wire model is not valid for applications if segment length to radius ratio smaller than about ten is required [15]. For the cases listed in Table 4.1, this ratio is equal to 21 and 10, respectively, meaning that it is still acceptable.

4.7. MONOPOLE ANTENNA

A monopole antenna is half of a dipole on a finite or infinite ground plane. The monopole antenna is usually fed from a coaxial cable through the ground plane. The monopole mesh can be created in a number of ways. In this chapter

MONOPOLE ANTENNA

we use the script monopole.m from the subdirectory mesh. This script allows us to create a monopole on a finite ground plane using a plate mesh, mouse input, and a strip mesh. Matlab function ginput is employed to acquire the mouse position. The script may also output an array of monopoles at arbitrary locations, monopoles on a bent plate, and so on (see Chapter 6).

As a first example, execute that script and do not change anything in the code. The result will be the Matlab picture shown in Fig. 4.8a. This figure shows the ground plane of the monopole antenna. Then follow the steps:

1. Use Matlab mouse cross and the left mouse button to mark white triangles shown in Fig. 4.8b. These triangles will define the feeding edge of the monopole (common edge of two triangles).
2. Press return key.

After pressing the return key the antenna structure is created. It appears on the screen as shown in Fig. 4.9a. The structure is a probe-fed monopole antenna of 1 m height on a 2×2 m ground plane. The strip width is 0.04 m. The strip has totally 7 rectangles and 14 triangles.

Looking through the script monopole.m, you will see that it employs the Matlab function delaunay in order to create the ground plane mesh. Then the probe feed is introduced as a common edge to the two triangles identified at step 1. The script monopole.m allows the following parameters to be changed

```
L=2.0;      %Plate length (along the x-axis)
W=2.0;      %Plate width (along the y-axis)
Nx=11;      %Discretization parameter (length)
Ny=11;      %Discretization parameter (width)
h=1.0;      %Monopole height
Number=7;   %Number of monopole rectangles
```

The script monopole.m introduces additional vertexes to refine the uniform mesh X, Y in the vicinity of the anticipated probe feed and control the monopole thickness. This is done by adding four points to the existing plate vertexes. The addition has the following form:

```
%Identify probe feed edge(s)
x=[-0.02 0.02];
y=[0 0];
X=[X x];
Y=[Y y];
C=mean(x);
x1=[C C];
y1=mean(y) +2*[max(x) -C min(x) -C];
X=[X x1];
Y=[Y y1];
```

68 DIPOLE AND MONOPOLE ANTENNAS: THE RADIATION ALGORITHM

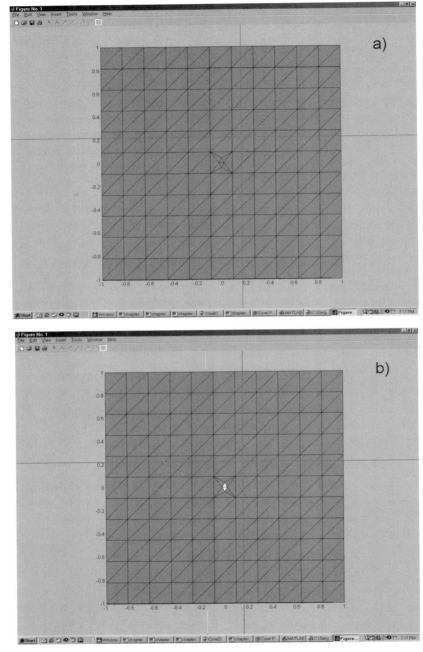

Figure 4.8. Two stages of monopole antenna creation. A Matlab mouse cross is seen.

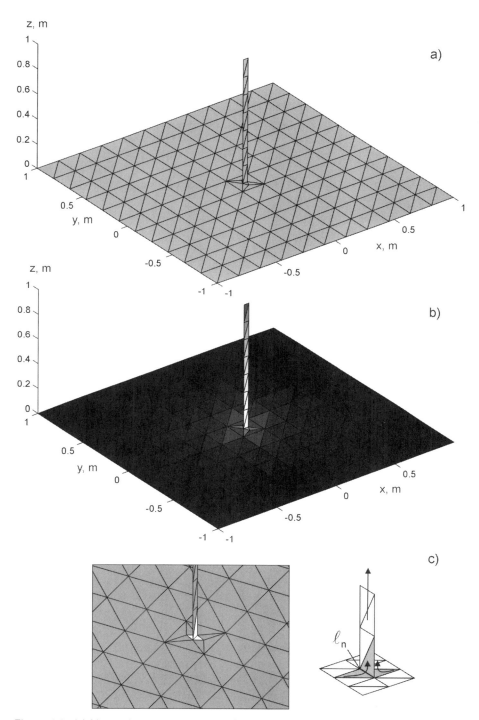

Figure 4.9. (a) Monopole antenna structure after execution `monopole.m`. (b) Surface current distribution at 75 MHz; the white color corresponds to larger magnitudes. (c) Model of the base-driven monopole.

Another point of interest is the triangle domain number, t(4,:). So far the entire antenna structure had the domain number 1. In the script monopole.mat, the domain number 0 corresponds to the plate triangles and the domain number 1– to the monopole triangles. One reason for that is the treatment of monopole–plate junction discussed below.

The script also enables multiple monopoles and multiple feeds. The monopole antenna structure is unlimited in size but it is recommended that the total number of triangles does not exceed 4000. The script can import the nonuniform plate meshes from the PDE toolbox and thus enhance the mesh quality and resolution close to the feed [16].

The edge at the junction (Fig. 4.9c) requires special consideration. There are three triangles attached to it. Therefore there are two RWG elements that correspond to the same edge. One of them includes the first triangle of the strip and the plate triangle to the left of the feed. Another includes the same strip triangle and the plate triangle to the right of the feed (Fig. 4.9c). Script rwg1.m of the present chapter automatically identifies such edges and creates two distinct RWG elements for every edge. The triangle domain number of the monopole and the plate must be different.

The antenna feeding edge is the junction edge (for the base-driven monopole). The model of the delta-function generator is applied, which is similar to that for the dipole. However, there are now two distinct RWG edge elements corresponding to the same edge. To separate these two elements, it is convenient to "double" the junction edge, that is, to repeat it twice in the mesh code. This is done in the script rwg1.m. The contribution of each single element should be taken into account. Equation (4.5) is transformed to

$$V_{n1} = l_{n1}V, \quad V_{n2} = l_{n2}V, \quad l_{n1} = l_{n2} \tag{4.9}$$

where indexes 1 and 2 refer to two distinct RWG junction elements. Equation (4.9) is employed in the script rwg4.m. Figure 4.9b shows the calculated surface current distribution at 75 MHz (script rwg5.m). Since the monopole height is chosen to be 1.0 m, we end up with a most popular combination, which is the quarter-wavelength monopole on a ground plane. The highest values of the surface current density are observed on the strip surface.

In the script rwg4.m, the default feeding edge was the one closest to the origin on the xy-plane. In the present case we have two such edges. This is in contrast to the dipole, where only one edge was used. Therefore the script rwg4.m introduces two different feed models: one for the dipole and another for the monopole

```
Index=INDEX  (1)    %Center feed - dipole
Index=INDEX  (1:2)  %Probe feed - monopole
```

where the array INDEX contains the edge numbers sorted against the edge distance from the origin. One should comment on either the first or second line of this code.

4.8. IMPEDANCE OF THE MONOPOLE

The formula for the input impedance of the dipole (4.8) should be extended to

$$Z_A = \frac{V}{l_{n1}I_{n1} + l_{n2}I_{n2}} \tag{4.10}$$

to account for the total current through the primary feeding edge. The antenna impedance calculation is again done in the script `rwg4.m`. The total current through the feed becomes

```
GapCurrent=sum(I(Index).*EdgeLength(Index)');
```

where `Index` is the array of the feeding edge numbers.

From a theoretical point of view, when a monopole is mounted on an ideally infinite ground plane, its impedance and radiation characteristics can be deduced from that of a dipole of twice its length in free space. For the base-driven monopole, the input impedance is equal to one-half that of the center-driven dipole. Also the monopole radiation pattern (Section 4.11) above the infinite ground plane is identical with the upper half of the radiation pattern of the corresponding dipole ([8], pp. 4–26). When the ground plane is of finite size, the impedance is not changed considerably but the radiation pattern is subject to change. The gain of the monopole is not equal to the dipole gain (Section 4.11).

Since the ground plane is finite, we are not supposed to have exactly one-half of the dipole impedance. However, the corresponding results appear to be similar. This can be seen in Table 4.2. The first column of Table 4.2 indicates the two different models used for comparison.

The SuperNEC model of the monopole on the finite ground plane used in Table 4.2 has $10 \times 10 = 100$ wire segments to model the 2×2 m plane. It is therefore probably too "transparent" to provide us with the accurate results.

Table 4.2. Input Impedance of Two Monopole Models at 75 MHz

Model	Input Impedance, Ω
Present calculation	$28 + j \times 22$
Monopole on a 2×2 m large wire-segment ground plane (SuperNEC)	$33 + j \times 32$

Note: The strip width is 0.04 m. The number of strip rectangles is 7. The equivalent wire radius is 0.01 m. The number of wire segments is 7. A WIPL-D simulation gives $27 + j \times 22 \, \Omega$.

Table 4.3. Radiated Electric Field of the Dipole at 75 MHz at the Distance of 100 m (Azimuthal Plane)

Model	E-field, V/m
Strip2.mat—244 triangles	0.0066 exp($-j*2.30$)
Wire with 39 segments (SuperNEC)	0.0062 exp($-j*2.26$)

4.9. RADIATION INTENSITY, RADIATED POWER, AND GAIN

The far-field parameters of an antenna were discussed in Chapter 3. Below we recall some of these definitions. The script efield1.m outputs the radiated electric/magnetic field at a point. The results for the radiated electric field of the 2 m long dipole at 100 m in the azimuthal plane are presented in Table 4.3.

As in Section 3.5 of Chapter 3, the radiation density of a field radiated by the antenna is given by the time-average Poynting vector [1, pp. 35–36]

$$\mathbf{W}(\mathbf{r}) = \frac{1}{2}\text{Re}[\mathbf{E}(\mathbf{r}) \times \mathbf{H}^*(\mathbf{r})] \quad (4.11)$$

which is expressed in W/m^2. Taking into account Eq. (3.4) of Chapter 3, we can write in the far field

$$\mathbf{W}(\mathbf{r}) = \frac{1}{2\eta}|\mathbf{E}(\mathbf{r})|^2 \frac{\mathbf{r}}{r} \quad (4.12)$$

The total power, radiated by the antenna in the far field can be written in the form

$$P_{rad} = \int_S \frac{1}{2\eta}|\mathbf{E}(\mathbf{r})|^2 dS \quad (4.13)$$

where S is a sphere surface surrounding the antenna. For Eq. (4.13) to hold, the sphere radius should be large enough compared to the antenna size and wavelength.

Matlab script efield2.m calculates the radiation density and radiation intensity of an antenna. We use the precise expression Eq. (4.11), which is valid in both far and near fields. More specifically, the norm of the Poynting vector is found. If necessary, that norm can be replaced by a projection of the Poynting vector onto the outer surface normal of the observation sphere.

The script also outputs the total power over a large sphere of a variable radius and the antenna gain (both logarithmic and linear), which are saved in the binary file gainpower.mat. The output of the script is shown in Fig. 4.10

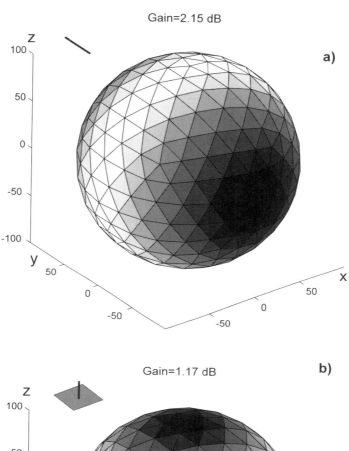
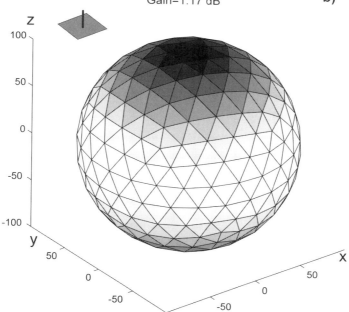

Figure 4.10. Radiation intensity over a sphere surface at 75 MHz. (a) Half-wavelength dipole along the y-axis (strip2.mat); (b) quarter-wavelength monopole along the z-axis (monopole.m). The white color corresponds to higher relative intensities.

for a half-wavelength dipole (`strip2.mat`) and a quarter-wavelength monopole (`monopole.m`).

The total radiated power of the dipole has a different value compared to the example of Chapter 3. This fact is to be expected, since the surface current magnitude due to 1 V antenna feed is obviously not equal to the magnitude created by the incidence of a 1 V/m plane wave. The total radiated power also slightly depends on the discretization accuracy.

However, for *all* four dipoles of Fig. 4.4, the linear gain is obtained as approximately 1.64 and the logarithmic gain is approximately $10\log_{10}(1.64) = 2.15\,\text{dB}$. These values exactly correspond to the theory data. The antenna gain and the total radiated power seem to be the most stable antenna parameter, weakly affected by various numerical approximations.

What is the gain of a quarter-wavelength monopole on a ground plane spanning infinitely in all directions? It might appear, at first sight, that it is equal to the gain of an equivalent dipole. However, this is not quite true. While the upper-half radiation pattern is exactly the same, the normalization factor U_0 in Eq. (3.9) of Chapter 3 is twice as small as the one for a similar dipole because there is no radiated power below the ground plane. Therefore the gain of the monopole is the dipole gain plus 3 dB [17, pp. 93–94]:

$$10\log_{10}(2) = 3\,\text{dB}$$

If the ground plane is finite, the gain may vary. In our case of the quarter-wavelength monopole on the half-wavelength plate, a very low gain of 1.2 dB is nevertheless obtained. This is rather an exception to the rule. When the plate width increases the gain increases as well.

For antenna elements with losses, the antenna gain includes one more important mechanism, namely ohmic losses in imperfect conductors. A more adequate definition of the gain will be considered in the following chapters.

4.10. RADIATION RESISTANCE AND DELIVERED ELECTRIC POWER

As far as the power transfer is concerned, an antenna can be treated as a resistive element much like the familiar resistor. The major difference is that the resistor dissipates electric power in the form of heat. The antenna dissipates electric power in the form of an electromagnetic signal radiated into free space. A total power of Joule's heat sources in a resistor R is given by

$$P = \frac{1}{2}R|I|^2 \tag{4.14}$$

Here I is the current amplitude, which is either real or complex. Since we are establishing an analogy between the antenna and the resistor, we assume that

Table 4.4. Numerically Found Values of Radiation Resistance versus Real Part of the Input Impedance; Numerically Found Values of Radiated and Delivered Power

Model	Real Part of the Input Impedance, Ω	Radiation Resistance, Ω	Radiated Power, W (P_{rad})	Feed Power, W (P_{feed})
Half-wavelength dipole at 75 MHz (strip2.mat)	87.6	88.2	0.0045	0.0044
Quarter-wavelength monopole at 75 MHz (monopole.m)	28.4	28.6	0.0109	0.0108

the total power, P_{rad}, radiated by the antenna has exactly the form of Eq. (4.14) where I is the current through the antenna feed. Simultaneously the resistance R is replaced by the "antenna" resistance, R_r, as follows:

$$P_{rad} = \frac{1}{2} R_r |I|^2 \qquad (4.15)$$

Equation (4.15) gives the radiation resistance, R_r, of the antenna in the form of Joule's law, Eq. (4.14). To find the radiation resistance, we must first calculate the total radiated power, P_{rad}. Then, using Eq. (4.15) together with the known value of the feed current, we solve for the radiation resistance. The radiation resistance is calculated in the script efield2.m. Table 4.4 compares the radiation resistance and the real part of the input impedance. Again, two examples considered are the half-wavelength dipole and a quarter-wavelength monopole (see Fig. 4.10).

From the table it can be seen that these values agree well with each other, to within a numerical round-off error. The reason is simple. The total power radiated by the antenna is also the total electric power delivered to the antenna in the feed:

$$P_{feed} = \frac{1}{2} \text{Re}(IV^*) = \frac{1}{2} \text{Re}(Z_A)|I|^2 = P_{rad} = \frac{1}{2} R_r |I|^2 \qquad (4.16)$$

Thus $R_r = \text{Re}(Z_A)$. Equation (4.16) provides another, and simpler, way to calculate the total power radiated by the antenna without the need of surface integration in the far field. The corresponding variable is FeedPower (script rwg4.m). It is not to be mixed with the variable TotalPower (script efield2.m). Calculation of the radiation resistance and the total power using two independent methods is the best way to test the correctness of the far-field results.

4.11. DIRECTIVITY PATTERNS

It was pointed out in Chapter 3 that the directivity pattern is the radiation intensity calculated over a circle in a certain plane, namely a cross section of the sphere shown in Fig. 4.10. More precisely, the antenna directivity is the normalized radiation intensity calculated in dB (Chapter 3, Section 3.7). Matlab script `efield3.m` calculates directivity patterns of the radiated field in the planes xy, xz, and yz, respectively. For a dipole along the y-axis (`strip2.mat`), we are interested in the yz-plane (elevation plane) and the xz-plane (azimuthal plane). Figures 4.11 and 4.12 show directivity patterns of two examples of the present chapter: a half-wavelength dipole at 75 MHz (`strip2.mat`) and a quarter-wavelength monopole at 75 MHz (`monopole.m`), respectively.

A problem is usually encountered when dealing with negative directivity values (in dB). Matlab polar plot is unable to handle these negative values. A way around this problem is to add an offset of, say, +40 dB to the directivity. The polar plot may then be rescaled as has been done in Figs. 4.11 and 4.12.

Note that in these figures the antenna gain is already calculated in the script `efield2.m`. If the radiation pattern includes the direction of maximum radiation, this will be simultaneously the pattern gain. Otherwise, the gain along a radiation pattern can be smaller than the overall antenna gain. Figure 4.12b provides an example. The script `efield3.m` can be modified to calculate the gain (`G=max(Polar_)`) for a given radiation pattern.

For the dipole (Fig. 4.11), the radiation patterns are essentially the same as those from Chapter 3. The antenna gain is 2.15 dB. On the dipole axis the directivity drops down to −62 dB. Note that the total sweep of 62 + 2 = 64 dB corresponds to the power ratio of

$$10^{64/10} \approx 2.5 \times 10^6$$

The patterns of a 1 m long monopole on the 2 × 2 m ground plane are given on the left of Fig. 4.12. The patterns of a 1 m long monopole on the 4 × 4 m ground plane are given on the right. These patterns are obtained if we type

```
L=4.0;   %Plate length (along the x-axis)
W=4.0;   %Plate width (along the y-axis)
```

in the script `monopole.m` and repeat the calculations for the monopole done in Sections 4.7 and 4.8.

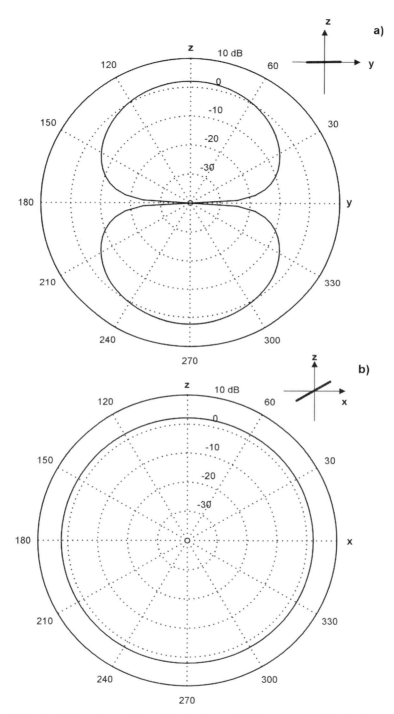

Figure 4.11. Directivity pattern of the radiated antenna field (efield3.m) of the half-wavelength dipole at 75 MHz (strip2.mat).

78 DIPOLE AND MONOPOLE ANTENNAS: THE RADIATION ALGORITHM

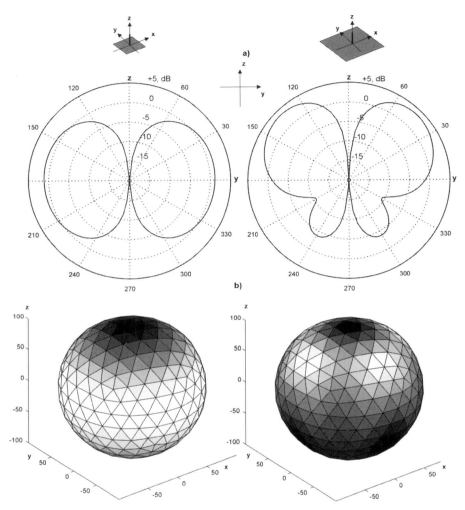

Figure 4.12. Directivity patterns and 3D radiation patterns of the radiated antenna field (efield3.m) of the quarter-wavelength monopole at 75 MHz. Left: 2 × 2 m ground plane; right: 4 × 4 m ground plane. OFFSET = 20 dB in the script efield2.m.

It is interesting to see from Fig. 4.12 how the wider ground plane effectively cuts the radiation below the horizon and makes the antenna more directional. The antenna gain increases from 1.2 dB to 3.7 dB if we double the size of the ground plane. The input impedance changes from $28 + j \times 22\,\Omega$ to $46 + j \times 16\,\Omega$, respectively. In the ideal case of the infinite ground plane, the monopole radiation pattern should be exactly the upper half of the dipole pattern. In every case (dipole or monopole) there is no radiation along the dipole/

monopole axis at elevation angle zero. According to Fig. 4.12 the monopole is also the (almost) omnidirectional antenna.

4.12. RECEIVING ANTENNA

In this section we compute an electric signal received by another antenna if the first antenna is the transmitter and the transmitting voltage amplitude is 1 V. The script efield1.m calculated electric and magnetic fields radiated by an antenna at any observation point in space using the current distribution on the antenna surface. The calculation method is equivalent to that discussed in Chapter 3. For the half-wavelength dipole strip2.mat at 75 MHz, the script outputs the electric field (in V/m)

```
EField=
-0.0000  -0.0000i
-0.0044  -0.0049i
-0.0000  -0.0000i
```

if the observation point is at 100 m, that is, ObservationPoint = [0;0;100].

Let us now place another (receiving) dipole at that observation point. The receiving antenna is none other than a scatterer, and the analysis of Chapter 2 can be applied to find the corresponding surface current distribution. We should therefore return to the Matlab directory of Chapter 2 to perform the necessary calculations.

To simplify file transfer, all necessary scripts from Chapter 2 are collected in subdirectory receivingantenna of the Matlab directory of Chapter 4. The input to rwg1.m is mesh strip2.m. The frequency is 75 MHz, and the incident electric field in script rwg4.m should be

```
kv=k*[0 0 1];
Pol=[0 -0.0044-j*0.0049 0];
```

where kv is the wave vector and Pol is the vector describing the direction, magnitude, and phase of the incident electric field. Figure 4.13 shows the surface current distribution of both radiating and receiving dipoles, respectively. The transmitter dipole calculated in this chapter is shown on top of the figure, whereas the receiver dipole is on the bottom. Both current patterns are equivalent to each other so we cannot distinguish between two dipoles.

After executing the command line

```
rwg1.m; rwg2.m; rwg3.m; rwg4.m; rwg5.m
```

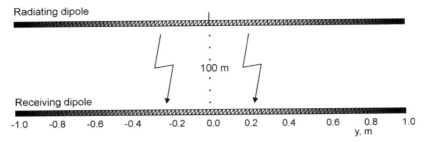

Figure 4.13. Magnitude of the surface current along strip strip2.mat. Top: radiating dipole antenna; bottom: receiving dipole antenna placed at 100 m from the transmitter and having the same orientation.

the surface current density through the middle edge of the receiving antenna is obtained as $-0.0045 - j0.0011$ A/m. Since the strip width is 0.02 m, the total current, I, in the feed (through the middle edge) is therefore $-0.000090 - j0.000023$ A. The same feed model is assumed for both receiving and transmitting antennas. The received voltage is given by

$$V_A = IZ_A \qquad (4.17)$$

where antenna impedance Z_A is found in Table 4.1. The voltage in the feed of the receiver is thus

$$V_A = -0.0068 - j0.0062 \, V$$

This complex number has the magnitude of approximately 9 mV and the phase of -2.4 rad. Worthy of note, these values should be compared to the voltage in the feed of the transmitting antenna that has the magnitude of 1 V and the phase factor zero. The radio link over 100 m thus reduces the voltage magnitude by approximately a factor of 100. Indeed, the present observation is valid only for the particular antenna size and particular frequency.

We have just outlined the method of finding both the magnitude (on the order of millivolts) and the phase of the receiving voltage at a given frequency. Receiving and transmitting antennas are assumed to be identical. This method can be applied to establish an antenna-to-antenna transfer function. The transfer function is obtained as the ratio of receiving to transmitting (1 V) voltage at different frequencies. The transfer function is important for nonsinusoidal antennas like ultra-wideband pulse antennas [18] since it allows us to predict the received pulse forms. This method will be used extensively in Chapter 9 where we consider a time domain ultra-wideband antenna.

4.13. FRIIS TRANSMISSION FORMULA

In conventional communication links with harmonic excitation, a simple and very effective power transmission formula is mostly used in the form (e.g., [1, p. 88] or [17, p. 61])

$$P_R = P_T \frac{G_T G_R}{(4\pi r)^2} \lambda^2 \tag{4.18}$$

where indexes R and T denote receiving and transmitting antennas, respectively, and G is the linear antenna gain. To apply this formula to the example of the previous section, we should choose $r = 100$ m, $G_T = G_R = 1.64$, $\lambda = 4$ m, and $P_T = 0.0044$ W (see Table 4.4) Therefore

$$P_R = 1.20 \times 10^{-7} \, W \tag{4.19}$$

The power transmission formula gives us the magnitude of the received power and voltage if the antenna impedance is known. However, the phase information is completely lost.

It is helpful to compare the power calculated using the Friis transmission formula with the power obtained using the direct scattering method from the previous section. In order to do so, we should consider a more realistic antenna circuit for the receiving antenna shown in Fig. 4.14. The circuit in Fig. 4.14 replaces the receiving antenna by its Thévenin equivalent: the ideal voltage source of strength V_A in series with Thévenin impedance, Z_A. Thévenin impedance is none other than the antenna input impedance. Thévenin voltage (received voltage) was already calculated in the previous section.

The power delivered to a load with impedance Z_L is given by (cf. Eq. (4.16))

$$P_L = \frac{1}{2}\text{Re}(Z_L)|I|^2 \tag{4.20}$$

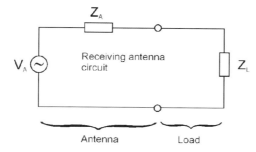

Figure 4.14. Equivalent circuit for the receiving antenna.

82 DIPOLE AND MONOPOLE ANTENNAS: THE RADIATION ALGORITHM

where I is the phasor of the current in the circuit in Fig. 4.14. Note that this current is not equal to the feed current calculated in Section 4.12 without the load (i.e., at $Z_L = 0$). We have

$$I = \frac{V_A}{Z_A + Z_L} \qquad (4.21)$$

Substitution of Eq. (4.21) into Eq. (4.20) gives

$$P_L = \frac{1}{2} \frac{\text{Re}(Z_L)}{|Z_A + Z_L|^2} |V_A|^2 \qquad (4.22)$$

When $Z_A = Z_L^*$, where the star denotes complex conjugate (i.e., a *conjugate matched load*), P_L attains its maximum value, which is given by [19, p. 618]

$$P_L = \frac{1}{8} \frac{|V_A|^2}{\text{Re}(Z_L)} \qquad (4.23)$$

Substituting $V_A = -0.0068 - j0.0062$ V (see the previous section) and $\text{Re}(Z_L) = 88\,\Omega$ yields

$$P_R = 1.20 \times 10^{-7}\, W \qquad (4.24)$$

This result comes in close proximity to the value obtained with Eq. (4.19) using the power transmission formula Eq. (4.18). It follows from there that Eq. (4.18) already takes into account the presence of a conjugate matched load in the antenna circuit.

4.14. CONCLUSIONS

There are dozens of variations on dipole and monopole antennas that have been developed over the years [1]. Some offer improvements in performance, while others are of interest for specialized applications. The dipole or a monopole is a basic element of antenna arrays considered in Chapter 6. The monopole antenna is actually more common than the dipole (just remember your car antenna).

From the numerical point of view, the monopole mesh on a finite or curved ground plane is more difficult to create than a simple dipole. A junction edge should be analyzed carefully, including two separate RWG elements sharing the same edge.

The algorithm of the present chapter is applicable to any metal antennas and will be utilized in the rest of the text. The useful changes to be made are

Matlab loops with regard to the radiation frequency and, possibly, with regard to other antenna parameters (antenna length, size, shaping, etc.). These loops are intended for antenna optimization purposes.

REFERENCES

1. C. A. Balanis. *Antenna Theory: Analysis and Design*, 2nd ed. Wiley, New York, 1997.
2. G. J. Burke and A. J. Poggio. *Numerical Electromagnetic Code NEC—Method of Moments. Part II. Program Description—Code*. Lawrence Livermore National Laboratory Report, UCID-18834, January 1981.
3. B. D. Popović, M. B. Dragović, and A. R. Djordjević. *Analysis and Synthesis of Wire Antennas*. Wiley, New York, 1982.
4. *MoM Technical Reference Manual*. Poynting Software Ltd., 2001.
5. R. F. Harrington. *Field Computation by Moment Methods*. Macmillan, New York, 1968.
6. J. W. Rockway, J. C. Logan, D. W. S. Tam, and S. T. Li. *The MiniNEC System: Microcomputer Analysis of Wire Antennas*. Artech House, Norwood, MA, 1988.
7. S. M. Rao, D. R. Wilton, and A. W. Glisson. Electromagnetic scattering by surfaces of arbitrary shape. *IEEE Trans. Antennas and Propagation*, 30 (3): 409–418, 1982.
8. R. C. Johnson, ed., *Antenna Engineering Handbook*, 3rd ed. McGraw-Hill, New York, 1993.
9. M. Davidovitz and Y. T. Lo. Rigorous analysis of a circular patch antenna excited by a microstrip transmission line. *IEEE Trans. Antennas and Propagation*, 37 (8): 949–958, 1989.
10. B. G. Salman and A. McCowen. The CFIE technique applied to finite-size planar and non-planar microstrip antenna. *Computation in Electromagnetics*, Conf. Publ. No. 420, 1996, pp. 338–341.
11. C. J. Leat, N. V. Shuley, and G. F. Stickley. Triangular-patch modeling of bow-tie antennas: Validation against Brown and Woodward. *IEE Proc. Microware Antennas Propagation*, 145 (6): 465–470, 1998.
12. S. R. Saunders. *Antennas and Propagation for Wireless Communication Systems*. Wiley, New York, 1999.
13. H. L. Bertoni. *Radio Propagation for Modern Wireless Systems*. Prentice Hall, Upper Saddle River, NJ, 1999.
14. R. W. P. King. *Tables of Antenna Characteristics*. IFI Plenum Data Corporation, New York, 1971, p. 39.
15. D. H. Werner. A method of moments approach for the efficient and accurate modeling of moderately thick cylindrical wire antennas. *IEEE Trans. Antennas and Propagation*, 46 (3): 373–382, 1998.
16. S. Makarov. MoM antenna simulations with Matlab: RWG basis functions. *IEEE Antennas and Propagation Magazine*, 43 (5): 100–107, 2001.
17. W. L. Stutzman and G. A. Thiele. *Antenna Theory and Design*. Wiley, New York, 1981.

84   DIPOLE AND MONOPOLE ANTENNAS: THE RADIATION ALGORITHM

18. T. P. Montoya and G. S. Smith. A study of pulse radiation from several broad-band loaded monopoles. *IEEE Trans. Antennas and Propagation*, 44 (8): 1172–1182, 1996.
19. K. R. Demarest. *Engineering Electromagnetics*. Prentice Hall, Upper Saddle River, NJ, 1998.

PROBLEMS

4.1. For a 2 m long dipole `strip2.mat` at 100 MHz determine:
  a. input impedance
  b. directivity patterns in the elevation and azimuthal planes
  c. antenna gain.

4.2. For a 2 m long dipole `strip2.mat` at 225 MHz determine:
  a. Radiation resistance and total radiated power
  b. Directivity patterns in the elevation and azimuthal planes
  c. Antenna gain.

4.3. Create a 1.0 m long monopole on a 2 by 2 m ground plane using the script `monopole.m`. The monopole frequency is 150 MHz. Determine:
  a. Input impedance and radiation resistance
  b. Directivity patterns in the elevation and azimuthal planes
  c. Antenna gain.

4.4.* Create a 0.2 m long monopole on a 2 by 2 m ground plane using the script `monopole.m`. The monopole frequency is 150 MHz. Determine:
  a. Input impedance and radiation resistance
  b. Directivity patterns in the elevation and azimuthal planes
  c. Antenna gain.

4.5. Script `multi1.m` in subdirectory `mesh` of the Matlab directory of Chapter 4 creates a horizontal dipole over a finite ground plane shown in Fig. 4.15a. This is the simplest type of a reflector antenna. Assuming a half-wavelength dipole, design the reflector, that is, find the optimal distance d from the plane in order to achieve the maximum antenna gain.

4.6. Solve Problem 4.5 under assumption that the total dipole length is equal to $3\lambda/2$.

4.7. Script `multi2.m` in the subdirectory `mesh` of the Matlab directory of Chapter 4 creates a horizontal dipole over a cylindrical metal reflector shown in Fig. 4.15b. This is another case of the reflector antenna. Assuming a half-wavelength dipole, design the reflector, that is, find the optimal distance reflector-dipole in order to achieve the maximum antenna gain.

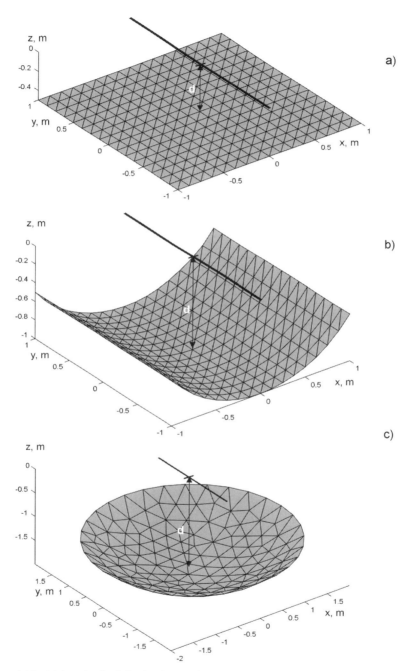

Figure 4.15. (a) Longitudinal dipole above a finite ground plane (multi1.m); (b) dipole with a cylinrical reflector (multi2.m); (c) dipole with a convex reflector (multi3.m). The feeding edge is at the origin.

86 DIPOLE AND MONOPOLE ANTENNAS: THE RADIATION ALGORITHM

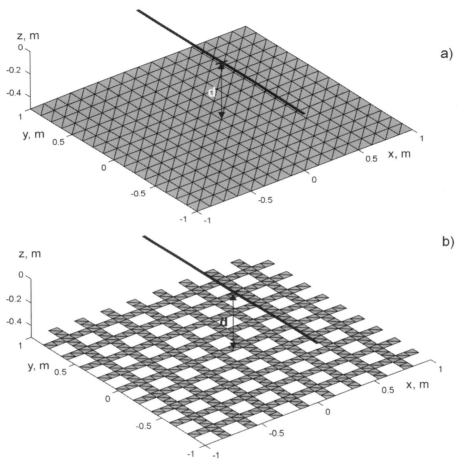

Figure 4.16. (a) Longitudinal dipole above a finite ground plane (multi1.m); (b) longitudinal dipole above a metal lattice (multi4.m). The feeding edge is at the origin.

4.8. Solve Problem 4.7 under assumption that the total dipole length is equal to $3\lambda/2$.

4.9. Script multi3.m in the subdirectory mesh of the Matlab directory of Chapter 4 creates a horizontal dipole over a convex metal reflector shown in Fig. 4.15c. This is one more case of a reflector antenna. Assuming a half-wavelength dipole, design the reflector, that is, find the optimal distance reflector-dipole in order to achieve the maximum antenna gain.

4.10. Solve Problem 4.9 under assumption that the total dipole length is equal to $3\lambda/2$.

Figure 4.17. Relative position of the receiving dipole.

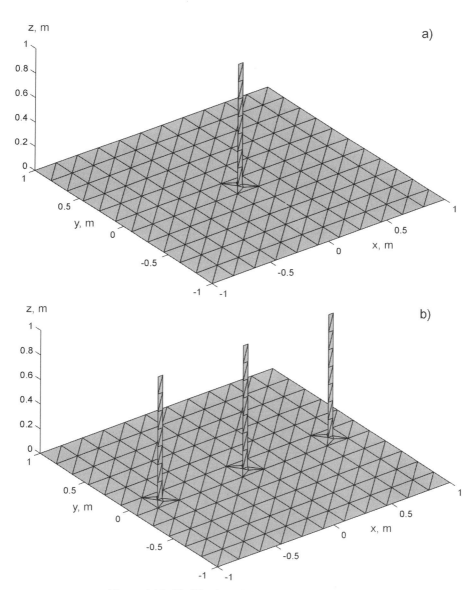

Figure 4.18. Modification of the script `monopole.m`.

4.11.* Script `multi4.m` in the subdirectory `mesh` of the Matlab directory of Chapter 4 creates a dipole over a planar metal lattice shown in Fig. 4.16b. The lattice was generated using the script `volumemesh1.m` in Appendix A. Which reflector works better: the solid surface in Fig. 4.16a or the metal lattice in Fig. 4.16b? Assume a half-wavelength dipole and $d = 0.5$ and 1.0 m, respectively.

4.12. A quarter-wavelength 1 m long monopole on a 2 by 2 m ground plane transmits a periodic signal at 75 MHz. The input voltage is 1 V. Determine voltage received by a half-wavelength dipole at the distance of 100 m from the monopole. The relative position of the dipole is shown in Fig. 4.17.

4.13.* In Problem 4.12 determine the total power received by the dipole antenna. Obtain the result using two different ways: Friis transmission formula Eq. (4.18) and the direct method outlined in Sections 4.12 and 4.13.

4.14.* The script `monopole.m` originally creates the monopole shown in Fig. 4.18a. Modify that script in order to create the combination shown in Fig 4.18b. Only the center monopole has the voltage feed. Two other monopoles are the passive elements. Plot the surface current distribution and the radiation intensity distribution over a large sphere for these two antenna types assuming the quarter-wavelength monopole. Is the second antenna more directional than the first?

5

LOOP ANTENNAS

5.1. Introduction
5.2. Loop Meshes and the Feeding Edge
5.3. Current Distribution of a Loop Antenna
5.4. Input Impedance of a Small Loop
5.5. Radiation Intensity of a Small Loop
5.6. Radiation Patterns of a Small Loop
5.7. Transition from Small to Large Loop: The Axial Radiator
5.8. Helical Antenna—Normal Mode
5.9. Helical Antenna—Axial Mode
5.10. Conclusions
References
Problems

5.1. INTRODUCTION

Wire loop antennas offer the advantages of low cost and low gain, and they are therefore useful in many portable wireless devices. Loop antennas replace the dipole-type radiators for applications where the receiver is held close to the body, in that the performance of a loop element is not degraded as much due to the high conductivity of the body [1]. Loop antennas take many different forms, including rectangle, square, triangle, ellipse, circle, among many others [2]. The circular loop is the most popular choice and has received the widest attention.

Small (compared to the wavelength) loop antennas are very poor radiators, and they are seldom employed for radio transmission. They are usually used in the receiving mode, such as in portable radios, where the antenna efficiency is not as important. They are widely used as probes for field measurement, including measurements within magnetic resonance imaging (MRI) coils.

Several references [3–7] indicate applicability of large-current rectangular loop antennas, with partial shielding of the loop path, to broadband and pulse transmission, in the nanosecond range. Ultra-wideband loop sensors for electromagnetic field measurements have recently been reported [8].

The task of the present chapter is to investigate loop antennas, including single-loop and helical antennas, with regard to the antenna impedance, directivity, and gain. The antenna analysis is essentially equivalent to that for the dipole and monopole (Chapter 4). We will pay considerable attention to electrically small and large loop antennas, and to helical antennas in the normal and axial mode of operation. For the purposes of comparison, the corresponding data for dipole antenna will be presented as well.

The antenna radiation algorithm (see Fig. 4.2 of the preceding chapter) is applied to loops without considerable changes. The code sequence is identical to that of Chapter 4. The corresponding Matlab scripts can be found in the Matlab directory of the present chapter. Before reading the chapter you may want to run these scripts. The final script displays the surface current distribution for a one-wavelength loop due to a feed voltage of 1 V.

For the flat structures similar to the thin-strip, planar bowtie antenna, or planar slot antenna, the Matlab PDE toolbox was quite adequate in the generation of triangular meshes (Chapter 2). For 3D looplike structures, we prefer to use relatively short custom Matlab scripts that allow us to easily create such complicated antenna shapes as a helical coil. The thin-strip model of Chapter 4 is applied to simulate a wire loop. This model already showed much promise in modeling thin dipoles.

5.2. LOOP MESHES AND THE FEEDING EDGE

Four scripts `loop1.m`, `loop2.m`, `loop3.m`, and `loop4.m` in subdirectory mesh of the Matlab directory of Chapter 5, create loop structures shown in Figs. 5.1 and 5.2, respectively. Before reading this chapter, you may want to run these scripts. The first script (`loop1.m`) creates a single closed loop of radius a. The strip width is h. The discretization accuracy along the loop circumference can be controlled depending on the anticipated frequency. Only one triangle is allowed across the strip.

The feeding edge of the loop antenna can be specified in different ways. We prefer using the (vertical) edge closest to the point [-1 0 0] in Cartesian coordinates. The feeding edge is indicated by a black bar in Fig. 5.1. This is in contrast to the dipole where the feeding edge is chosen by the one closest to the origin. Changes, related to the proper identification of the feeding edge, are done in the script `rwg4.m`. Then, the delta gap model of Chapter 4 is applied to the feeding edge.

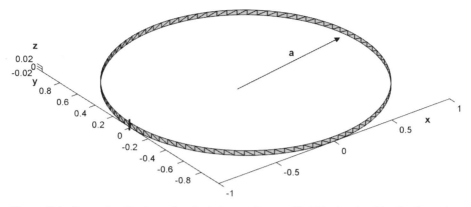

Figure 5.1. Geometry structure of a single-loop antenna with 180 triangles. The feeding edge is marked by a black bar.

The script loop2.m (or loop3.m) creates a helical coil shown in Fig. 5.2a. The coil radius, turn spacing, and the number of turns can be varied, along with the width of the strip. The feeding edge is chosen as the one closest to the point [-1 0 0] in Cartesian coordinates. The feeding edge is indicated by a black bar in Fig. 5.2a. The difference between scripts loop2.m and loop3.m is that the script loop3.m keeps vertical edges perpendicular to the strip direction, even for a helical coil. The script loop4.m creates a conical helix coil (a tapered helix) shown in Fig. 5.2b. The coil parameters can be varied, along with the width of the strip. The feeding edge is chosen as the one closest to the point $[-a_{min}\ 0\ 0]$ in Cartesian coordinates, where a_{min} is the center radius of the coil. The feeding edge is indicated by a black bar in Fig. 5.2b.

According to the image theory [2], the dipole-like coil with 12 turns in Fig. 5.2a approximately corresponds to a monopole-like coil with 6 turns over the infinite ground plane. The ground plane coincides with the xy-plane. Analogously, the conical helix in Fig. 5.2b can be used to simulate a monopole helix of 5 turns over the infinite perfectly conducting ground plane. Also a finite ground plane can be introduced in a way similar to the method of Chapter 4 for the monopole. The scripts loop1.m-loop4.m are organized nearly in the same manner. We introduce the vertex points and divide a bent strip into equal rectangles. Then each of these rectangles is divided into two right triangles. No Delaunay triangulation (delaunay.m) is employed to create loop meshes.

5.3. CURRENT DISTRIBUTION OF A LOOP ANTENNA

Using the script loop.1m in the subdirectory mesh, we create a loop of the radius $a = 1$ m and the width 0.04 m. The loop contains 180 triangles. The loop circumference is $C = 2\pi a = 6.28$ m. Then, the main code sequence

92 LOOP ANTENNAS

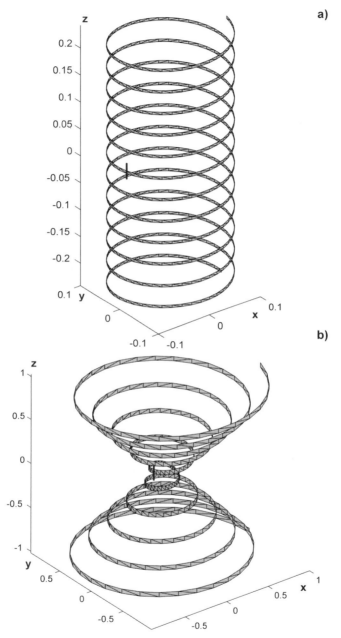

Figure 5.2. (a) Helical coil with 12 turns, turn spacing 0.04 m, and strip width 0.005 m; (b) helical tapered coil with 10 turns, center radius of 0.1 m, end radius of 1 m, and strip width 0.05 m. Feeding edge is marked by a black bar.

rwg1; rwg2; rwg3; rwg4; rwg5;

is executed, where loop1.mat should be given as the input file to rwg1.m. Script rwg5.m outputs not only the resulting current distribution but also the maximum and minimum electric current along the loop. These values are obtained as the maximum (minimum) current density magnitude multiplied by the width of the strip (the length of the feeding edge). Note that the script rwg5.m is slightly changed to acquire the number of the feeding edge from rwg4.m.

The important parameter for the loop antenna is the ratio of the loop circumference, $C = 2\pi a$, and the radiated wavelength, λ. At very low frequencies, when $C \ll \lambda$, a model of *electrically small loop* is considered. When the circumference is comparable to the wavelength, $C \approx \lambda$, a model of *electrically large loop* should be used. These models have very different radiation and impedance properties. Let us examine first a very low frequency of 0.48 MHz (specified in rwg3.m), when $C = 0.01\lambda$. At that frequency, the current distribution along the loop is practically uniform. Script rwg5.m outputs the minimum current (at the antenna feed) of 57.09 mA and the maximum current (the opposite side of the loop) of 57.12 mA. Note that these values are much larger in magnitude than the current of a comparable dipole antenna at low frequencies. This is why the loop is sometimes called the *current radiator* [6,7], in contrast to the dipole, which is in that sense the *voltage radiator*.

The uniformity of the current distribution along the loop constitutes the background of the analytical model for the small loop [2, pp. 217–221]. We will demonstrate that this assumption is becoming inaccurate very quickly, when the wavelength becomes smaller than approximately 10 loop circumferences. Figure 5.3 shows the computational results for the surface current when $C = 0.1\lambda$, $C = 0.5\lambda$, $C = \lambda$, and $C = 2\lambda$. This corresponds to frequencies 4.8 MHz, 23.9 MHz, 47.7 MHz, and 95.5 MHz, respectively, if C is fixed as 6.28 m.

We recall that the color bar always extends from minimum to maximum current magnitude. Only the loop with $C = 0.1\lambda$, where $i_{min} = 5.48$ mA and $i_{max} = 5.80$ mA, indicates a relatively uniform current distribution. The loop with $C = 0.5\lambda$ already predicts a very deep minimum of the surface current close to the voltage feed. The loop with $C = \lambda$ has two maxima of the current distribution along the loop, whereas the loop with $C = 2\lambda$ has four. The situation resembles the corresponding result for the dipole (Fig. 4.6 of Chapter 4) where a periodic pattern of the surface current develops along the antenna at higher frequencies.

One can see that when $C > 0.1\lambda$, the current along the loop is actually not as high in magnitude and is comparable to the current of the dipole antenna, under the assumption of the same voltage in the antenna feed.

The realistic (nonsymmetric) current patterns indicate that the realistic loop cannot have zero radiation along its axis. This is in contrast to the simplified theory for the loop, which assumes a constant current along the circumference [2, pp. 217–221].

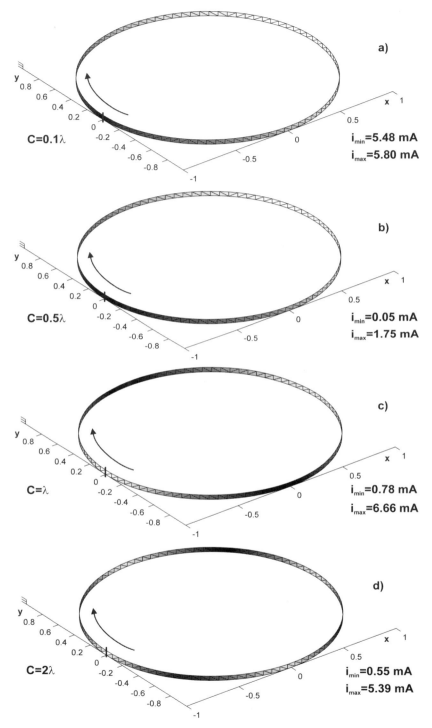

Figure 5.3. Surface current distribution and maximum/minimum current along the loop at different frequencies.

One can test the surface current distribution of a loop at different values of the strip width (diameter of an "equivalent wire"). The corresponding calculations have been done and the results similar to Fig. 5.3 were obtained when the strip width varies from 0.005 to 0.05 m. Somewhat larger current magnitudes are observed for "thicker" loops. Note that according to Eq. (4.1) of Chapter 4, the strip width of 1 cm corresponds to the wire radius of 2.5 mm.

Calculations of this and the following sections may be rescaled to geometrically smaller loops. For example, the current distribution of Fig. 5.3c simultaneously corresponds to a wire loop of 20 cm in diameter, with the wire diameter of 2 mm at the frequency of 477 MHz.

5.4. INPUT IMPEDANCE OF A SMALL LOOP

The real part of the input impedance (the radiation resistance) characterizes the power radiated by an antenna if we assume a lossless metallic antenna structure. Since we are interested here in the radiation properties of the loop at a given feed voltage, the input impedance should be investigated first. A loop from the previous section, with the radius of 1 m and the strip width $h = 0.04$ m, is again chosen. For the purposes of comparison, we also investigate a vertical dipole, whose length is equal to the loop circumference, that is, to 6.28 m. The dipole is modeled by a thin strip of the same width. The dipole mesh, oriented along the z-axis is saved in the binary file `strip.mat`.

Table 5.1 shows real and imaginary parts of the input impedance for several selected frequencies, in the low-frequency range. The results are obtained using scripts `rwg3.m`, `rwg4.m`. At low frequencies the loop has a large positive complex impedance (*inductance*) and vanishingly small radiation resistance. The loss resistance of single-turn loop is, in general, much larger than its radiation resistance. Therefore the corresponding radiation efficiency (ratio of the radiated power to the power dissipated into heat on the loss resistance) of the loop is very low and depends on its loss resistance. The dipole at low frequencies has a very large negative complex impedance (*capacitance*) and a small, but finite, radiation resistance. This resistance is 100 to 10 times greater

Table 5.1. Input Impedances of Loop and Dipole Antennas at Low Frequencies (`loop1.mat` and `strip.mat`)

λ	f, MHz	Loop		Dipole of length = C	
		$\text{Re}(Z_A)$, Ω	$\text{Im}(Z_A)$, Ω	$\text{Re}(Z_A)$, Ω	$\text{Im}(Z_A)$, Ω
$C = \lambda/10$	4.8	0.02	183	1.8	−1646
$C = \lambda/8$	6.0	0.06	235	2.8	−1289
$C = \lambda/6$	8.0	0.22	331	5.1	−922
$C = \lambda/4$	11.9	1.78	594	12.3	−533

96 LOOP ANTENNAS

Table 5.2. Power Delivered to Loop and Dipole Antennas by a 1 V Feed Voltage at Low Frequencies (`loop1.mat` and `strip.mat`)

C	f, MHz	Power, Loop μW	Power, Dipole μW
$C = \lambda/10$	4.8	0.34	0.33
$C = \lambda/8$	6.0	0.54	0.84
$C = \lambda/6$	8.0	1.0	3.0
$C = \lambda/4$	11.9	2.5	21.7

than the loop resistance. Therefore the dipole presents a more significant resistance to the generator circuit even at very low radiating frequencies.

Script `rwg4.m` outputs the power delivered to the antenna in the feed in the form

$$P_{\text{feed}} = \frac{1}{2}\text{Re}(I_{\text{gap}} \cdot V^*_{\text{gap}}) \qquad (5.1)$$

where I_{gap} is the gap (feed) current and V^*_{gap} is the gap voltage. The star denotes complex conjugate. In our case, $V^*_{\text{gap}} = 1\,\text{V}$. For a lossless antenna, Eq. (5.1) simultaneously gives the total radiated power. Table 5.2 outlines power obtained in this way for the loop and for the dipole, respectively. The same frequencies as before are used. One can see that, at low frequencies, the dipole outputs more power than the loop assuming other equal conditions (the ideal voltage generator). At the same time the radiated power of the loop at ultra-low frequencies may be higher than the dipole power as indicated by the second row of Table 5.2. This is why the small loop antenna may be an attractive choice for superconducting circuits, where the loss resistance is exactly zero [9–11].

5.5. RADIATION INTENSITY OF A SMALL LOOP

Matlab script `efield2.m` calculates the radiation intensity distribution over a large sphere. Figure 5.4 shows the computational results for two examples of the previous section: the loop with the radius of 1 m and a vertical dipole, whose length is equal to the loop circumference, that is, to 6.28 m. The frequency is 4.8 MHz, which corresponds to $C = 0.1\lambda$.

It might appear at first sight that the small loop and the small vertical dipole radiate very similarly. Both antennas are omnidirectional and have the maximum directivity in the azimuthal plane. The major difference is connected, however, to the polarization of the radiated field. Whereas for the vertical dipole in Fig. 5.4b the radiated electric field is vertically polarized, the

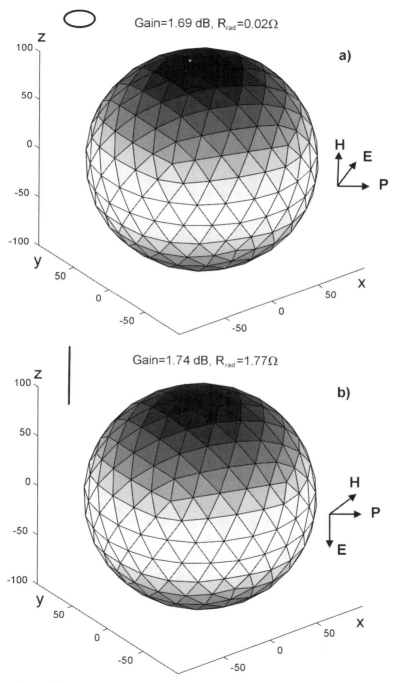

Figure 5.4. Radiation intensity distribution: (a) Small loop with circumference $C = 0.1\lambda$; (b) small dipole of equivalent length. The frequency is 4.8 MHz.

loop in Fig. 5.4a possesses the horizontal polarization of the electric field. The change in polarization of the electric field does not affect the direction of the Poynting vector, **P**, in Fig. 5.4, which is always directed outward.

A combination of the dipole and a loop constitutes the so-called electric-magnetic dipole, which is the omnidirectional antenna with a circular polarization [2, pp. 512–513].

To check the polarization directions, we can apply the script efield1.m that calculates the E- and H-fields at a point. We choose the observation point as a p = [100; 0; 0]m. This gives

```
E(loop) = 1.0e-004 ×           E(dipole) = 1.0e-004 ×
  0.0000 + 0.0000i               0.0000 + 0.0000i
  0.4456 + 0.3296i               0.0000 + 0.0000i
  0.0000 + 0.0000i              -0.3983 + 0.3630i
```

in accordance with the directions shown in Fig. 5.4.

Many analytical formulas for small loops are presented in Ref. [2, ch. 5]. In particular, in the far field [2, p. 213], one has in terms of spherical coordinates r, θ, ϕ,

$$E_\phi = V_0 \frac{e^{-jkr}}{r} \sin\theta, \quad E_r = E_\theta = 0; \quad V_0 = \frac{\eta \pi S i_0}{\lambda^2} \tag{5.2}$$

$$H_\theta = -\frac{V_0}{\eta} \frac{e^{-jkr}}{r} \sin\theta, \quad H_r = H_\phi = 0 \tag{5.3}$$

where the z-axis is the polar axis. Here $r = |\mathbf{r}|$; $\eta = \sqrt{\mu/\varepsilon} \approx 377\,\Omega$ is the free-space impedance; $S = \pi a^2$; and i_0 is a constant current around the loop. In our case (see Fig. 5.3a), $i_0 \approx 5.64$ mA and $f = 4.8$ MHz; $\lambda = 62.5$ m. This gives $V_0 \approx 5.37$ mV. Substitution of this value into Eq. (5.2) at $r = 100$ m yields the theoretical magnitude of the E-field as $53.7\,\mu$V/m. The corresponding numerical prediction is $55.4\,\mu$V/m and agrees very well with the simplified theory. However, in contrast to the magnitude, the constant-current model does not match the numerically calculated phase values.

5.6. RADIATION PATTERNS OF A SMALL LOOP

Script efield3.m gives radiation patterns in the planes xy, xz, and yz, respectively. We are mainly interested in the yz-plane (elevation plane). Figure 5.5 compares the radiation patterns for the loop and the dipole, respectively. Again, the loop with the radius of 1 m and a vertical dipole, whose length is equal to the loop's circumference, are considered. The frequency is 4.8 MHz, which corresponds to $C = 0.1\lambda$.

RADIATION PATTERNS OF A SMALL LOOP 99

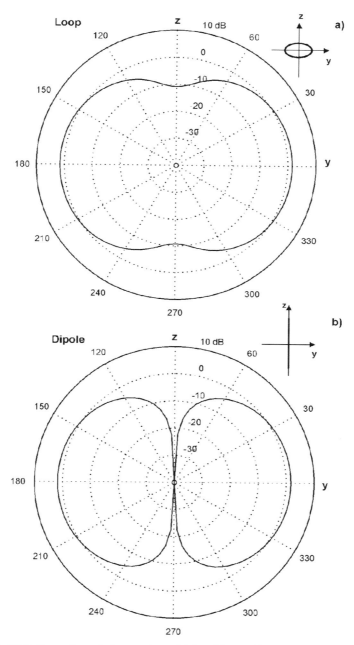

Figure 5.5. Radiation patterns in the yz-plane: (a) Small loop with circumference $C = 0.1\lambda$; (b) small dipole of equivalent length. The frequency is 4.8 MHz.

The major difference between the dipole and the small, but finite, loop is the nonzero radiation intensity in the axial direction. If the electric current along the loop is assumed to be a constant, then the radiation intensity on the loop axis should be exactly zero, due to symmetry reason. Even a small nonuniformity in the surface current causes a considerable distortion of the radiation pattern in the axial direction so that the "theoretical" null in the axial direction disappears. In particular, for the radiation pattern shown in Fig. 5.5a, the current distortion is obtained as low as 5.5% (see Fig. 5.3a).

To obtain a deep minimum in the axial direction, the operation frequency needs to be chosen extremely low. Also the discretization of the loop should be refined in order to avoid numerical inaccuracy.

5.7. TRANSITION FROM SMALL TO LARGE LOOP: THE AXIAL RADIATOR

The loop has a remarkable property consisting in that its radiation pattern transforms from a "broadside" type to the "end-fire" type when the radiation frequency increases. To demonstrate such a transition, let us consider first the case where the loop circumference is a fraction of wavelength, namely $C = 0.1\lambda$, $C = 0.2\lambda$, and $C = 0.3\lambda$, respectively. We run the sequence

```
rwg3;  rwg4;  efield2.m;  efield3m;
```

for the each of these frequencies, which are 4.8 MHz, 9.5 MHz, and 14.3 MHz, respectively. C is 6.28 m for the loop with the radius of 1 m. Figure 5.6a shows the radiation patterns in the H-plane. To be consistent with Ref. [2], the radiation patterns are rescaled in this section versus maximum directivity. It is seen how the minimum in the axial direction of the loop disappears with increasing frequency and the antenna pattern becomes more "symmetric."

It is expected that if the frequency continues to increase, the maximum radiation direction is shifted toward the radial direction. To support this conclusion, we present in Fig. 5.6b three radiation patterns corresponding to three higher frequencies so that $C = 2\pi\lambda/10 \approx 0.63\lambda$, $C = 2\pi\lambda/5 \approx 1.26\lambda$, and $C = 2\pi\lambda/2 \approx 3.14\lambda$. The corresponding frequencies are 30 MHz, 60 MHz, and 150 MHz, respectively, for the loop with the radius of 1 m.

It is seen that, when the circumference of the loop is about one wavelength, the maximum radiation is along the loop axis, which is perpendicular to the plane of the loop. Because of its many applications, the one-wavelength circumference circular loop antenna is considered as fundamental as a half-wavelength dipole [2, p. 221]. Any further increase in frequency leads to sidelobes as shown in Fig. 5.6b and is not very useful.

The input impedance of the electrically large loop can be calculated as a function of frequency using the method of Chapter 7 below. Here we only check the impedance at $C = \lambda$ (at 48 MHz for the loop with the radius

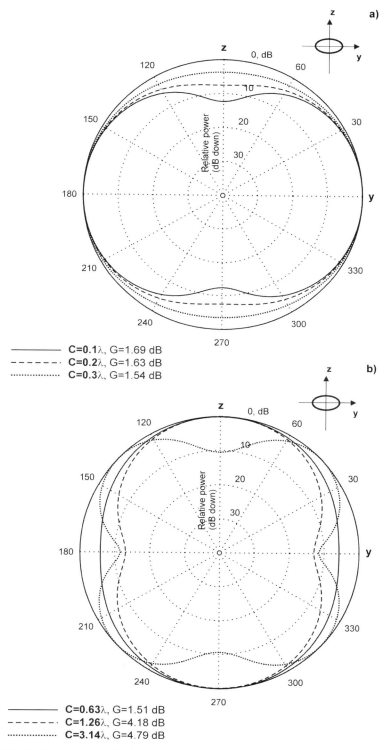

Figure 5.6. Radiation patterns in the yz-plane: (a) Electrically small loop; (b) electrically large loop. G is overall gain (found using `efield2.m`).

of 1 m). Script rwg4.m outputs $Z_A = 116 - j88\,\Omega$, which is a very reasonable value. For analytical results related to large loops and recent references see [12,13].

Finally in this section we note that Fig. 5.6b is created in such a way to exactly meet the conditions of Fig. 5.6 in Ref. [2, p. 220]. In [2], however, a model of a constant-current loop is employed to obtain the corresponding radiation patterns analytically. This model is expected to be very inaccurate at these relatively high frequencies. The disagreement between theory and simulations appears to be very significant in the case of Fig. 5.6, including incorrect shapes of the radiation patterns. The analytical curves in Fig. 5.6 of Ref. [2] should therefore be replaced by a more reliable result.

5.8. HELICAL ANTENNA—NORMAL MODE

A helical antenna shown in Fig. 5.2a is a combination of several loops. The helical antenna is a basic, simple, and practical configuration of an electromagnetic radiator. The calculation for the helical antenna is done exactly in the same way as for the single loop. Script loop2.m in the subdirectory mesh creates a mesh loop2.mat for a helical antenna shown in Fig. 5.2a. Loop radius a, turn spacing S, number of turns N, and the strip width h need to be given for the helical antenna. The binary file loop2.mat should be specified as an input to the script rwg1.m in directory Matlab.

The proper identification of the feeding edge for the helical antenna (script rwg4.m) may constitute a problem. The feeding edge should be the one perpendicular to the strip (a vertical edge) and should be located in the middle of the antenna. This means a center-fed helix or a helical antenna above the infinite perfectly conducting ground plane. The feeding edge is indicated by a black bar in Fig. 5.2a. One solution is to choose the feeding edge closest to the point [-1 0 0] or, maybe, to another point [-a 0 0] in Cartesian coordinates. This method works well for tested examples if the number of discretization rectangles is even. Use p(:,Edge_(:,Index)) to check the position of the feed edge.

First, we investigate the performance of the helical antenna in the so-called *normal mode* [2, pp. 505–508]. In the normal mode of operation, the field radiated by the antenna is maximum in a plane normal to the helix axis and minimum along its axis (similar to the electrically small loop). To achieve the normal mode of operation, the dimensions of the helix are usually very small compared to the wavelength. More precisely, the entire length of the helix should be small compared to the wavelength as well [2, p. 508].

Using the script loop2.m from the subdirectory mesh, a helical antenna of $N = 9$ turns is created, with radius $a = 10$ cm, and loop spacing $S = 4$ cm. The width of the strip is 5 mm (the equivalent wire radius is 1.25 mm) and the total height of the helix is 36 cm. The number of triangles per turn is 80.

HELICAL ANTENNA—NORMAL MODE

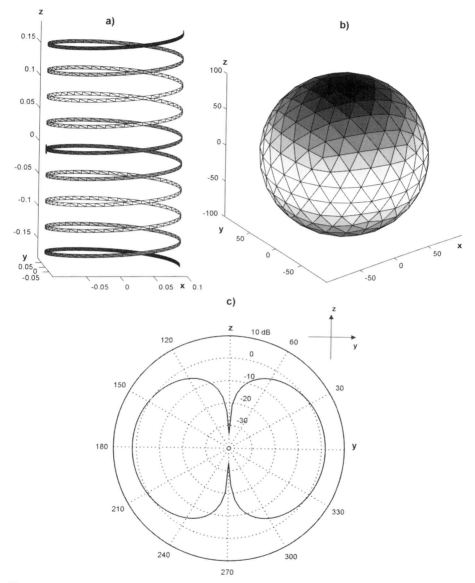

Figure 5.7. (a) Surface current distribution (rwg5.m); (b) radiation intensity distribution (efield2.m); and (c) the radiation pattern (efield3.m) of the center-fed 9-turn normal-mode helical antenna at 40 MHz.

Figure 5.7 shows the surface current distribution, the radiation intensity distribution, and the radiation pattern (yz-plane) of the center-fed helix at 40 MHz. This frequency corresponds to the ratio $C/\lambda = 0.084 < 0.1$. The total length of the wire is

$$L_n = N\sqrt{C^2 + S^2} = 5.7\,\text{m} \tag{5.4}$$

The wire length is smaller than the wavelength of 7.5 m at 40 MHz. The condition of the normal mode is thus satisfied to certain extend.

The major advantage of the helical antenna in the normal mode over a single loop is a large increase in the input resistance. For the antenna considered in this section, script `rwg4.m` outputs the input resistance of 51 Ω. This value provides an almost ideal match to the 50 Ω transmission line and is nearly 5000 (!) times higher than the impedance of a single loop of the same diameter. Since the size of the helical antenna is on the same order, this antenna is clearly superior to the small loop.

Theory for electrically small loops predicts that a helix with N loops has the input resistance N^2 times the input impedance of the single loops [2, p. 209]. For the present antenna this factor is even higher than N^2. The reason is perhaps that the present antenna is approaching the resonance as the entire structure. In Fig. 5.7a the corresponding current macrostructure is seen along the wire length. Note that theory, which is based on the simple addition of many loops with equal current distribution [2], is unable to describe that effect.

Thus, even though the radiation resistance of a single-turn loop may be small, the overall value can be significantly increased by including many turns and employing the helical antenna instead of the loop. This is a very desirable and practical mechanism that is not available for the small dipole. Similar to the single loop, the helical antenna in the normal mode is the omnidirectional antenna.

The normal mode of operation is very narrow in bandwidth and its design is rather tricky, because of the critical dependence of its radiation characteristics on its geometrical dimensions [2, p. 508]. In Table 5.3 we present the input impedance of the helical antenna from this section as a function of frequency. Table 5.3 explains why the frequency of 40 MHz was chosen as a center frequency of the helical antenna. The results of Table 5.3 will change

Table 5.3. Input Impedances of the 9-Turn Normal-Mode Helical Antenna with the Radius of 10 cm and the Total Height of 36 cm

C/λ	f, MHz	$\text{Re}(Z_A)$, Ω
0.059	28	0.5
0.071	34	2.1
0.080	38	10.6
0.084	**40**	**50.6**
0.088	42	4267.0
0.101	48	5.5
0.122	58	1.0

drastically if we change antenna geometry (increase or decrease turn spacing, number of turns, etc.). Problems at the end of this chapter provide the corresponding examples.

5.9. HELICAL ANTENNA—AXIAL MODE

In the *axial mode*, the helical antenna rather resembles an end-fire array of loops (see the next chapter). To excite this mode, the diameter of the loop and the loop spacing must be large fractions of the wavelength [2, p. 509]. When the loop spacing is large, the strip mesh will be considerably deformed in the vertical direction. Therefore, instead of the script loop2.m, we will use the script loop3.m, which keeps the orthogonality of transversal edges to the instantaneous strip direction. Comparative study of the two mesh types has revealed, however, that the mesh's "orthogonalization" does not have a significant effect on the computational results. The difference in input impedance is typically several percent if we keep the same discretization accuracy.

We consider here an example that essentially reproduces the shape of a helical antenna reported in [15, pp. 13–16/7]. Using the script loop3.m from the subdirectory mesh, a helical antenna of $N = 15$ turns is created, with $a = 5.45$ cm and loop spacing $S = 7.6$ cm. The width of the strip is 5 mm, and the total height of the helix is 114 cm (57 cm for a helical antenna on the infinite ground plane). The number of triangles per turn is 24 (the number of rectangles is 12). The total number of triangular structures is 358 (357 RWG edge elements). Such a rough discretization is primarily used to speed up the computation process. The problems at the end of this chapter address the issue of mesh refinement.

Figure 5.8 shows the surface current distribution, the radiation intensity distribution, and the radiation pattern (elevation plane) of the center-fed helix at 635 MHz. This frequency corresponds to the ratio $C/\lambda = 0.72$. The total length of the wire is

$$L_n = N\sqrt{C^2 + S^2} = 5.26 \text{ m} \tag{5.5}$$

The wire length is thus more then 10 times the wavelength of 47 cm at 635 MHz.

A big advantage of the helical antenna in the axial mode over the single loop is the large increase in antenna directivity. For the antenna considered in this section, the script efield2.m outputs a gain of 10.28 dB, which is more than 6 dB (!) higher than the gain of the single loop at the same frequency. The increase in the directivity results from the fact that the helical antenna in the normal mode is actually an array of the co-radiated loops, each of which contributes into directivity. More sophisticated arrays of independently driven antenna elements, considered in the next chapter, operate essentially in the same way.

106 LOOP ANTENNAS

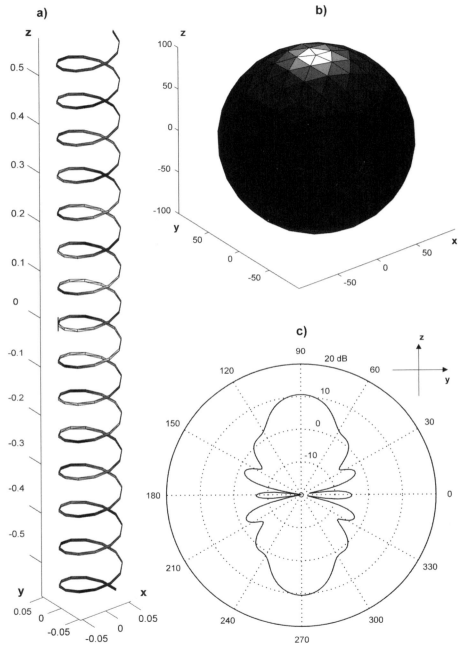

Figure 5.8. Surface current distribution (rwg5.m), radiation intensity distribution (efield2.m), and the y^2-plane radiation pattern (efield3.m) of the center-fed 15-turn axial-mode helical antenna at 635 MHz.

The antenna's impedance is calculated in the script rwg4.m. The result is $Z_A = 44 + j \times 23\,\Omega$, with an inductive imaginary part. The impedance value can be refined with an increase in the mesh's grid size. The total radiated power is 9.0 mW. This is higher than the value of the half-wavelength dipole in Chapter 4.

Another very inviting property of the helical antenna in the axial mode is the almost *circular polarization* of the field in the main lobe. For example, at the point [0; 0; 100] m on the antenna axis, the script efield1.m outputs the *E*-field in the form

```
E   =
  0.0171 + 0.0071i
 -0.0075 + 0.0150i
  0.0000 + 0.0000i
```

in V/m. The *x*- and *y*-components of the electric field appear to be very close in magnitude.

It has already been shown that the axial mode of operation can be generated with great ease [2, p. 509]. The circular polarization of the main lobe can be achieved by setting $C \approx \lambda$ and $S \approx \lambda/4$, where *S* is the loop spacing [2, p. 509]. These results, adopted from [15], are related to a helix with a large number of turns (10 and more), backed by a circular cavity of a finite size. The center-fed helix is different from this configuration, because of a higher sensitivity of input impedance and a gain to frequency. Even when the helix circumference is larger than two-thirds wavelength, large variations of impedance and gain are observed. Table 5.4 shows the simulation results for the gain of the 15-turn helix considered in this section. Very similar gain values are observed if we double the number of boundary elements. Figure 5.9 presents another set of antenna simulation results that demonstrate how the performance changes with frequency, at the frequency of 500 MHz. For a comprehensive study of helical antennas in the axial mode, see [15,16].

Table 5.4. Gain of the 15-Turn Axial-Mode Center-Fed Helical Antenna

C/λ	f, MHz	G, dB
0.517	500	6.7
0.628	550	4.9
0.685	600	5.6
0.725	**635**	**10.3**
0.742	650	10.0
0.799	700	4.5
0.856	750	3.3
0.913	800	3.2
0.970	850	3.8

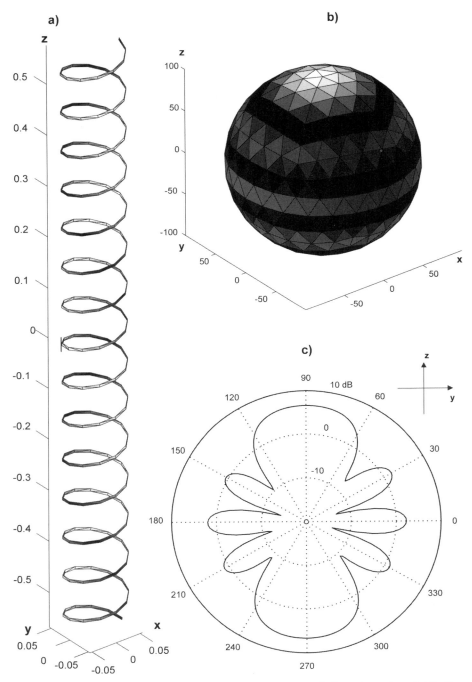

Figure 5.9. Surface current distribution (rwg5.m), radiation intensity distribution (efield2.m), and the y^2-plane radiation pattern (efield3.m) of the center-fed 15-turn axial-mode helical antenna at 500 MHz.

5.10. CONCLUSIONS

In this chapter we applied the antenna radiation algorithm of Chapter 4 to the study of loop antennas. The major points of interest include antenna input impedance, radiation intensity distribution, gain, and radiation patterns. Electrically small and large loop antennas, as well as the helical antenna in the normal and axial mode of operation were considered.

The helical antenna in the axial mode provided us with the first example of an antenna array. Single loops are combined in the helix to improve performance. Normally such a combination makes the antenna more directional and considerably increases the gain (helix in the axial mode). Other antenna parameters can be altered as well, such as the antenna impedance (helix in the normal mode). In the next chapter we will consider antenna arrays of separate antenna elements, also with independent feeds.

In addition to the types studied in this chapter the family of loop antennas includes such important members as half-loop antennas over a ground plane and shielded loop antennas [17].

REFERENCES

1. K. Fujimoto and J. R. James, eds. *Mobile Antenna System Handbook*. Artech House, Dedham, MA, 1994.
2. C. A. Balanis. *Antenna Theory. Analysis and Design*, 2nd ed. Wiley, New York, 1997.
3. H. F. Harmut. *Antennas and Waveguides for Nonsinusoidal Waves*. Academic Press, New York, 1984, ch. 2.
4. H. F. Harmut and S. Ding-Rong. Antennas for nonsinusoidal waves. I Radiators. *IEEE Trans. Electromagnetic Compatibility*, 25 (1): 13–24, 1983.
5. H. F. Harmut and S. Ding-Rong. Antennas for nonsinusoidal waves. II Sensors. *IEEE Trans. Electromagnetic Compatibility*, 25 (2): 107–115, 1983.
6. H. F. Harmut and N. J. Mohamed. Large current radiators. *IEE Proceedings H*, 139 (4): 358–362, 1992.
7. G. P. Pochanin. Large current radiator for the short electromagnetic pulses radiation. In E. Heyman, B. Mandelbaum, and J. Shiloh, eds., *Ultra-wideband, Short-Pulse Electromagnetics 4*. Kluwer Academic/Plenum Publishers, New York, 1999, pp. 149–155.
8. A. Yarovoy, R. de Jongh, and L. Ligthart. Ultra-wideband sensor for electromagnetic field measurements in time domain. *Electronics Letters*, 36 (20): 1679–1680, 2000.
9. R. C. Hansen. Superconducting antennas. *IEEE Trans. Aerospace Electronic Systems*, 26 (2): 345–355, 1990.
10. R. J. Dinger, D. R. Bowling, and A. M. Matrin. A survey of possible passive antenna applications of high-temperature superconductors. *IEEE Trans. Microwave Theory Techniques*, 39 (9): 1498–1507, 1991.

11. G. G. Cook and S. K. Khamas. Control of radar cross sections of electrically small superconducting antenna elements using a magnetic field. *IEEE Trans. Antennas and Propagation*, 42 (6): 888–890, 1994.
12. L. Le-Weii, L. Mook-Seng, K. Pang-Shyan, and Y. Tat-Soon. Exact solutions of electromagnetic fields in both near and far zones radiated by thin circular-loop antennas: A general representation. *IEEE Trans. Antennas and Propagation*, 45 (12):1741–1748, 1997.
13. L.-W. Li, C.-P. Lim, and M.-S. Leong. Method-of-moments analysis of electrically large circular loop antennas: Nonuniform currents. *IEE Proc. Microwave Antennas Propagation*, 146 (6): 416–420, 1999.
14. S. K. Khamas, G. G. Cook, S. P. Kingsley, R. C. Woods, and N. M. Alford. Investigation of the enhanced efficiencies of small superconducting antenna elements. *Journal of Applied Physics*, 74 (4): 2914–2918, 1993.
15. H. E. King and J. L. Wong. Helical antennas. In R. C. Johnson, ed., *Antenna Engineering Handbook*, 3rd ed. McGraw-Hill, New York, 1993, ch. 13.
16. H. Nakano. *Helical and Spiral Antennas*. Wiley, New York, 1987.
17. G. S. Smith. Loop antennas. In R. C. Johnson, ed., *Antenna Engineering Handbook*, 3rd ed. McGraw-Hill, New York, 1993, ch. 5.

PROBLEMS

5.1. A loop of radius $a = 1$ m (structure loop1.mat in subdirectory mesh) is investigated at the frequencies corresponding to $C = \lambda/1000$ and $C = \lambda/100$, where C is loop circumference. Determine:

   a. Input impedance
   b. Directivity pattern in the yz-plane
   c. Antenna gain.

5.2. A loop of radius $a = 1$ m (structure loop1.mat in subdirectory mesh) and a dipole of length $L = C = 6.28$ m are investigated at frequencies corresponding to $C = \lambda/1000$ and $C = \lambda/100$. Determine:

   a. Power delivered to the loop by 1 V voltage source in the feed
   b. Power delivered to the dipole by 1 V voltage source in the feed.

5.3. The analytical formula for the input resistance of a small loop is given by $\text{Re}(Z_A) = 20\pi^2(C/\lambda)^4$ [2, p. 209], where C is loop circumference. Compare input resistance values of a small loop from Table 5.1 with the corresponding values obtained using this formula.

5.4.* The analytical formula for the total radiated power of a small loop is given by $P_{\text{rad}} = \eta(\pi/12)(ka)^4|i_0|^2$ [2, p. 209], where η is the free-space impedance and k is the wavenumber. Compare values obtained using this formula with the values from Table 5.2. Hint: Repeat calculations from Table 5.2, and compute i_0 (mean current around the loop) using script rwg5.m.

5.5. Create a loop of 20 cm in diameter with 180 triangles (90 associated rectangles in script `loop1.m`). The width of the strip is 5 mm (the equivalent wire radius is 1.25 mm). Assuming a frequency corresponding to one wavelength circumference ($C = \lambda$), determine:
   a. Input impedance
   b. Antenna radiation pattern in the yz-plane
   c. Antenna gain
   d. Power delivered to the loop by 1 V voltage feed.

5.6. Solve Problem 5.5 for the case where the loop is replaced by a dipole of length $L = C = 0.628$ m.

5.7. Investigate the effect of the mesh grid size (the number of triangles in the boundary element loop structure) on the input impedance of a loop at $C = \lambda$, where C is loop circumference. Consider 80, 180, and 360 triangles (40, 90, and 180 associated rectangles in script `loop1.m`).

5.8. Investigate the effect of the strip width on the input impedance of a loop at $C = \lambda$, where C is loop circumference. Consider the loop of 20 cm in diameter with $h = [2:2:16]$ mm, and 180 triangles (90 rectangles in script `loop1.m`).

5.9. Create a loop of 10 cm in diameter, with the strip width of 2 mm. The resonant frequencies of the loop are given by $C = n\lambda$, $n = 1, 2, 3, \ldots$ [17]. Calculate the input impedance of the loop at three lowest resonant frequencies.

5.10. Create a wire loop of 2 cm in diameter, with an equivalent wire radius of 0.5 mm. The resonant frequencies of the loop are given by $C = n\lambda$, $n = 1, 2, 3, \ldots$ [17], where C is loop circumference. Calculate the input impedance of the loop at three lowest resonant frequencies.

5.11. Obtain:
   a. Input impedance
   b. Radiation intensity distribution
   c. Radiation patterns (the elevation plane)
   of an electrically large loop at $C = 3\lambda$, $C = 6\lambda$, and $C = 9\lambda$. Use a loop structure with 360 triangles (180 associated rectangles in the script `loop1.m`).

5.12. The helical antenna of Section 5.8 operates in the normal mode. Update the values of the input resistance from Table 5.3 if the loop spacing changes from 4 cm to 3 cm. Keep all other antenna parameters the same.

5.13. The helical antenna of Section 5.8 operates in the normal mode. Update the values of the input resistance from Table 5.3 if the number of turns increases from 9 to 11. Keep all other antenna parameters the same.

5.14.* Design an omnidirectional helical antenna operating in the normal mode at a center frequency of 500 MHz with the input resistance of approximately 50 Ω. Plot the antenna radiation pattern in the yz-plane.

5.15. The helical antenna of section 5.9 operates in the axial mode at 635 MHz. Increase the number of rectangles to 24 per turn (number of triangles is 48). Obtain the antenna input impedance, and compare that value with the one given in Section 5.9. Repeat the problem for 48 rectangles.

5.16. The helical antenna of Section 5.9 operates in the axial mode at 635 MHz. Obtain:

    a. Antenna gain

    b. Radiation intensity distribution

when the number of turns, N, takes values 5, 11, and 13.

6

ANTENNA ARRAYS: THE PARAMETER SWEEP

6.1. Introduction
6.2. Array Generators: Linear and Circular Arrays
6.3. Array Terminal Impedance
6.4. Impedance and Radiated Power of Two-Element Array
6.5. How to Organize the Matlab Loop
6.6. Array Network Equations
6.7. Directivity Control
6.8. Broadside Array
6.9. End-Fire Array
6.10. Pattern Multiplication Theorem
6.11. Comparison of Theory and Simulation
6.12. Optimization of End-Fire Array: The Phase Loop
6.13. Hansen-Woodyard Model
6.14. Power Map of End-Fire Array
6.15. Phased (Scanning) Array
6.16. Array of Bowties over Ground Plane
6.17. On the Size of the Impedance Matrix
6.18. Conclusions
References
Problems

6.1. INTRODUCTION

The inertia of a rotating radar antenna means that the time constant of a typical installation is not likely to be less than a second. This results in a loss of precious time, if the antenna is to be pointed at several targets in succession. A system that orients the antenna beam without inertia, on the other hand, allows all the possibilities of radar to be used in the full. It is in this way that the beam of a surveillance radar, scanning the horizon, can pause for an instant to confirm or disprove a possible alert. Similarly a tracking radar with electronic scanning can follow several targets by pointing at each of them successfully without any loss of time.

There is a process, known for many years, by which an electronic scanning beam can be obtained: such systems are known as *phased arrays* [1,2]. The method consists of feeding power to a number of radiating simple antennas (the array) by means of phase shifters such that the phase progressions along the array follow a characteristic with an arithmetic progression. The array thus produces a narrow beam whose orientation depends precisely on these phase shifts. This type of antenna has been produced since the late 1960s, when sufficiently reliable phase shifters first appeared [2].

The antenna algorithm developed in Chapters 4 and 5 can be extended, in a straightforward manner, to the antenna arrays. The new point is the multiple antenna feed. The number of feeding edges is now equal to the number of antennas in an array. Each feeding edge is characterized by its own (phase-shifted) voltage in the feed. We will later see that the input impedance of a single antenna in an array will be different from that in a free space, due to mutual coupling between array elements.

Successive numerical steps necessary to calculate an antenna array are implemented in the scripts `rwg1.m` - `rwg5.m`. These scripts may be found in the Matlab directory of the present chapter. Before reading the chapter, you may want to run these scripts. The final script displays the surface current distribution for an array of two half-wavelength dipoles. After the first code sequence is complete, scripts `efield2.m` and `efield3.m` provide radiation patterns of the antenna including 3D patterns and the antenna gain. The calculation of the input impedances is done in the script `rwg4.m`. The radiation resistance is calculated in the script `efield2.m`. The present code sequence is applicable to arbitrary arrays, and not only to the arrays of dipoles or monopoles. An example of a practical array calculation is considered at the end of the chapter.

Antenna arrays can be designed to control their radiation characteristics by properly selecting the phase/amplitude distribution and spacing between the elements. The corresponding design process implies a loop with regard to one or more parameters, such as the array spacing. The conversion of the Matlab scripts to a loop will be considered in this chapter. The corresponding codes are collected in subdirectories `loop1` and `loop2` of the Matlab directory of Chapter 6.

6.2. ARRAY GENERATORS: LINEAR AND CIRCULAR ARRAYS

The script multilinear.mat in the subdirectory mesh of the Matlab directory of Chapter 6 makes it possible to create a linear array containing an arbitrary number of dipoles (a realistic number shouldn't exceed 20!). Run this script first to see the result. The corresponding dipole mesh is given in the binary file strip1.mat (with 80 triangles). The dipole length is 2 m and the strip width is 0.02 m (the equivalent wire radius is 0.005 m). These parameters are optional and can be changed if necessary. Also, another dipole structure can be used as an input (see Chapter 4).

To create a dipole array, the mesh cloning algorithm is employed. It is based on the following Matlab code:

```
p=[p pbase];
t=[t tbase+length(pbase)];
```

which combines two arbitrary nonintercepting antenna structures described by p, t and pbase, tbase, respectively, into a single mesh p, t. The structures under study can be a dipole and a ground plane, or two dipoles, or etc. We will use that code to combine together many identical antenna meshes (e.g., dipoles), shifted in space, in order to create an antenna array.

The code multilinear.m outputs the array mesh into the binary file array.mat. The same is valid for other array generators, such as multicircular.m, multimonopole.m, and multibowtie.m considered in this chapter. The file array.mat is used as an input to rwg1.m. The array in the script multilinear.m is created in such a way that all the dipoles are directed along the z-axis. Figure 6.1 shows an array of eight dipoles that is created using multilinear.m and displayed using function viewer array. The dipole separation distance, d, is 1 m. In order to position the array in space, we follow the Kraus's book [1, p. 281]. The array axis is the x-axis and the vertical direction is the z-direction. Note that this geometry is different from dipole orientation in Chapter 4 where the dipole axis is the y-axis.

All array generators output an array Feed(1:3,1:N), where N is the number of array elements. This array identifies the feeding edge positions for each array element. In contrast to the last two chapters, the array Feed is an important part of the code sequence. This array is saved in binary file mesh2.mat, which is used as an input to rwg3.m.

An alternative to the planar dipole array shown in Fig. 6.1 is the circular array shown in Fig. 6.2a. The circular array consists of identical parallel dipoles equally spaced around the circumference of a circle. Only center-fed dipoles are used as array elements, and the center of each dipole is located on the circumference of the circle, in the xy-plane. The coordinate system used is similar to that for the planar array. The array "axis" is the x-axis and the

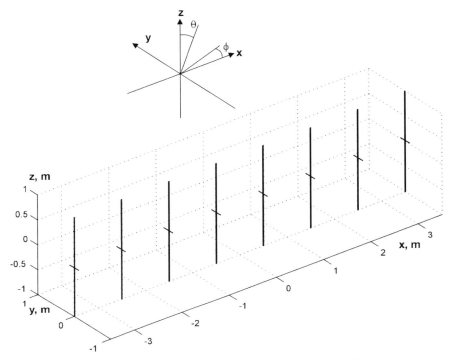

Figure 6.1. Linear array of eight dipoles created by miltilinear.m. Dipole feeding edges (shown by a bar) are in the xy-plane.

vertical direction is the z-direction. The circular array is an array configuration of very practical interest. Its applications span radio direction finding, air and space navigation, underground propagation, radar, sonar, and many other systems. A comprehensive theory of circular arrays is given in [3] (see also [4, ch. 6]). The script multicircular.m in the subdirectory mesh of the Matlab directory of Chapter 6, creates a circular array with an arbitrary number of dipoles (a realistic number shouldn't exceed 20) over a circle of arbitrary size.

The circular array of monopoles (Fig. 6.2b) is created using the script multimonopole.m. This script is very similar to the script monopole.m from Chapter 4. To create an array of base-driven monopoles, we again use the mouse input and mark pairs of triangles with the common edge. We repeat step 1 of Section 4.7 several times to create several monopoles at arbitrary positions. After that, the return key is pressed to fix the result.

In contrast to the script monopole.m, no mesh refinement is made close to the monopole feeds. The feeding edges are just the plate edges. To obtain the skinner strips, we have to refine the entire plate mesh (increase Nx and Ny in multimonopole.m).

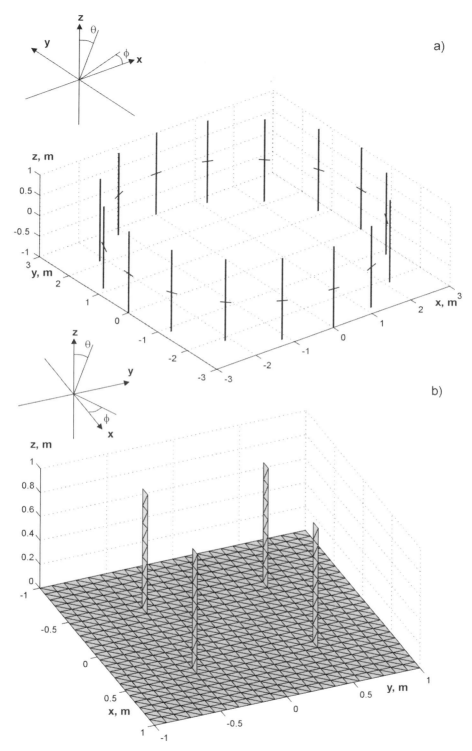

Figure 6.2. (a) Circular array of 16 dipoles created by `milticircular.m`; dipole feeding edges are in the xy-plane. (b) Circular array of four base-driven monopoles created by `miltimonopole.m`.

6.3. ARRAY TERMINAL IMPEDANCE

As far as array impedance is concerned, the advantage of the numerical approach is the ability of the direct calculation of an important coupling characteristic: the terminal or driving impedance of each array element. The input impedance, Z_n^{in}, of a single antenna n is defined in Chapter 4 (Section 4.6) in the form

$$Z_n^{in} = \frac{V^n}{J^n} \tag{6.1}$$

where V^n is the feed voltage and $J^n = l_n I_n$ is the corresponding total current through the feeding edge. Because of the electromagnetic interaction between different array elements, Eq. (6.1) cannot be automatically applied to the antenna array. Other array elements irradiate element n and thus change the feed current J^n. Therefore the value of Z_n^{in} has to be changed as well.

For an array of N elements, the *terminal impedance*, Z_n^{in}, of the array element n is given by Eq. (6.1) under the assumption that all the array elements are simultaneously excited by the corresponding voltages $V^1, \ldots, V^N$ [1, pp. 282–283]. Words "driving" impedance [3, p. 426] or "active" impedance [5, p. 155] are also used.

This circumstance offers a way to calculate the terminal array impedances using the approach developed previously for a single antenna. As in Chapters 4 and 5, the impedance calculations are done in the script rwg4.m. A few modifications need to be made in that script. First we find driving edges for each array element as those closest to the given array Feed:

```
N=length(Feed(1,:));
Index=[];
for k=1:N
    for m=1:EdgesTotal
        n1=Edge_(1,m);
        n2=Edge_(2,m);
        Distance(m)=norm((p(1:3,n1)+p(1:3,n2))/2- ...
        Feed(1:3,k));
    end
    [Y,INDEX]=sort(Distance);
    Index=[Index INDEX(1)];    %Center feed - dipole
end
```

The array Index outputs the numbers of driving edges. The length of this array equals N, namely the number of array elements. We must assume here that the driving voltages are all equal to 1 V and all have the same zero phase, that is,

```
V(Index)=1.0*EdgeLength(Index);
```

Further the system of moment equations is solved using Gaussian elimination. The terminal impedances are then found by applying Eq. (6.1) to each element of the array separately. The corresponding loop, which also calculates the power radiated by the array elements (lossless arrays only), is written in the form

```
for n=1:N
    nn=Index(n);
    GapCurrent(n)=I(nn)*EdgeLength(nn);
    GapVoltage(n)=V(nn)/EdgeLength(nn);
    Impedance  (n)=GapVoltage(n)/GapCurrent(n);
    FeedPower(n)=1/2*real(GapCurrent(n)* . . .
    conj(GapVoltage(n)));
end
```

Total radiated power is simply the sum of powers radiated by each antenna element.

For example, the terminal impedance of two half-wavelength dipoles (strip1.mat) at 75 MHz, separated by 2 m (by a half wavelength) becomes (run the code sequence rwg1.m - rwg4.m and do not change anything in the code)

$$Z_{1,2}^{in} = 67 + j \times 13\Omega \tag{6.2}$$

The value predicted in Eq. (6.2) is close to the early estimation found in Kraus [1, p. 283] who predicted $60 + j \times 14 \Omega$. The total antenna array impedance at a driving point is given by terminal impedances in parallel [1], namely (see Fig. 6.3)

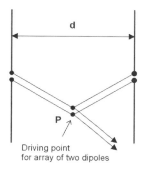

Figure 6.3. Array of two dipoles with arrangement for driving elements.

$$Z_P^{in} = \frac{1}{1/Z_1^{in} + 1/Z_2^{in}} = 34 + j \times 7\,\Omega \qquad (6.3)$$

The same formula for the total impedance is valid for an array with an arbitrary number of elements.

6.4. IMPEDANCE AND RADIATED POWER OF TWO-ELEMENT ARRAY

It is interesting to check the behavior of the terminal impedance as a function of array spacing, d. For simplicity we again consider the array of two identical parallel half-wavelength dipoles with the length of 2 m and width 0.02 m (strip2.mat). Figure 6.4 shows the real (input resistance) and imaginary (input reactance) part of the terminal impedance for each array element.[1]

Since there is no difference in relative positions of the two array elements, both of them clearly have the same terminal impedance. It is seen that when the distance between array elements increases, the terminal impedance approaches the impedance of the corresponding half-wavelength dipole (Chapter 4, Table 4.1) shown in Fig. 6.4 by a dashed line.

Another quantity of interest is the total radiated power of the array. We consider the same example of two dipoles. The total power is found using the formula [1, p. 284]

$$P_r = P_r^1 + P_r^2 = \frac{1}{2}\text{Re}(I^1 V^{1*}) + \frac{1}{2}\text{Re}(I^2 V^{2*})$$
$$= \frac{1}{2}\text{Re}(Z_1^{in})|I^1|^2 + \frac{1}{2}\text{Re}(Z_2^{in})|I^2|^2 \qquad (6.4)$$

The star denotes complex conjugate. Equation (6.4) assumes an array of lossless antenna elements. Figure 6.5 shows the total radiated power as a function of the separation distance, d. The dashed line at the bottom is the power radiated by a single half-wavelength dipole (0.0045 W; see Table 4.4 of Chapter 4). The upper dashed line is twice that value. When the distance between the array elements increases, the total radiated power clearly tends to the value of 0.0090 W. However, at relatively small separation distances, the total array power may be as much as 60% higher. The first power resonance occurs when the separation distance is approximately 2 m (half-wavelength). At that point the terminal impedance is given by Eq. (6.2). It is predominantly real and matches 50 Ω better than the impedance of the single half-wavelength dipole (Chapter 4). This inviting array property appears for mutually-coupled array elements.

---

[1] Although this example may be simple from the numerical point of view, its theoretical description is very complicated.

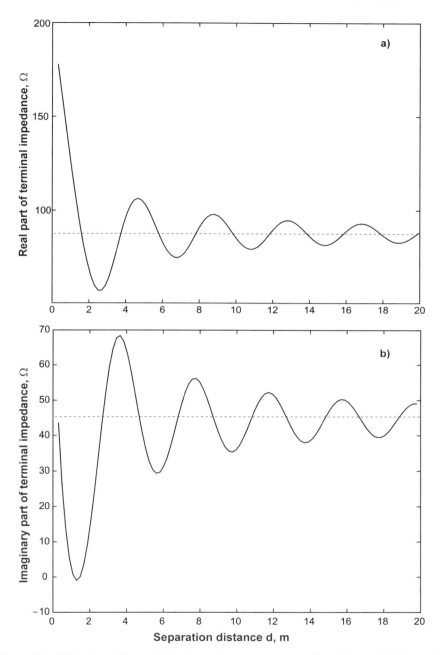

Figure 6.4. (a) Input resistance; (b) input reactance versus separation distance, d, for two half-wavelength dipoles, 2 m long each. The dashed line shows the corresponding result for a single dipole.

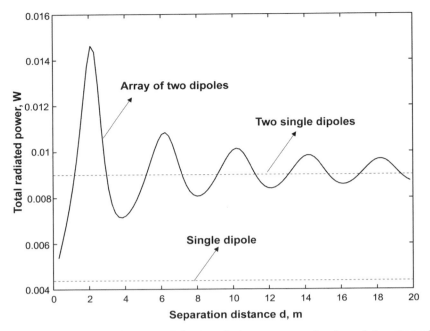

Figure 6.5. Total radiated power of the two-dipole array as a function of the separation distance, d. Dashed lines show the corresponding results for a single dipole and for two uncoupled dipoles, respectively.

6.5. HOW TO ORGANIZE THE MATLAB LOOP

The results of the previous section were obtained by organizing the source codes multilinear.m, rwg1.m - rwg4.m into a loop with respect to the variable separation distance, d. The corresponding loop codes are saved in the subdirectory loop1. The first step is to convert the scripts into Matlab functions by adding a function line at the top of them. For example, the script rwg1.m is converted to the function by adding the line

function []=rwg1

in place of the clear all statement at the top of the script. The script multilinear.m needs an input argument, d, which is the antenna separation distance. Therefore the following line is invoked:

function []=multilinear(d)

The script rwg4.m needs two output arguments. They are the array of terminal impedances and the total power, that is,

```
function [Impedance, PowerSum]=rwg4
```

The next step is to create a wrapper script containing the loop itself. This script (loop.m) yields the anticipated loop in the following form:

```
for k=1:M
    d(k)=dS + (k - 0.5*Step);
    multilinear(d(k));
    rwg1; rwg2; rwg3;
    [Impedance(k,:), PowerSum(k)]=rwg4;
end
```

You may want to run the script loop.m first to estimate the processor time necessary to perform the loop with $M = 100$. The script output was given in Figs. 6.4 and 6.5, respectively. The already precalculated data are saved in the binary file loop.mat in the same subdirectory. Load this file and plot PowerSum versus d, that is, (plot(d,PowerSum); grid on;) in order to obtain Fig. 6.5.

6.6. ARRAY NETWORK EQUATIONS

From the viewpoint of the transmission line theory [6,7], the relationships between the feed currents and voltages in the antenna elements are given by the so-called network equations:

$$V^1 = Z_{11}J^1 + Z_{12}J^2 + \ldots + Z_{1N}J^N$$
$$V^2 = Z_{21}J^1 + Z_{22}J^2 + \ldots + Z_{2N}J^N$$
$$\ldots$$
$$V^N = Z_{N1}J^1 + Z_{N2}J^2 + \ldots + Z_{NN}J^N \quad (6.5)$$

where Z_{mn} defines the feed voltage at antenna m due to a unity current in antenna element n when the current in all the other feeds is zero. When indexes m and n are not identical, the term Z_m represents the *mutual impedance* between elements m and n. The term Z_{mm} is called the *self-impedance* of element m.

The terminal impedances introduced above can be expressed in terms of self- and mutual impedances. For example,

$$Z_1^{in} = \frac{V^1}{J^1} = Z_{11} + \frac{J^2}{J^1}Z_{12} + \ldots + \frac{J^N}{J^1}Z_{1N} \quad (6.6)$$

Not only Z_{mn} but also the feed currents are necessary to find the terminal impedance using this approach. Equations (6.5) and (6.6) are useful when

124 ANTENNA ARRAYS: THE PARAMETER SWEEP

analytical approximations are available for the feed currents, similar to the point source approximation. These approximations [4, pp. 426–428] will not be considered in the present text. From a numerical point of view, Eqs. (6.5) and (6.6) are rather inconvenient, since we would have to disconnect the corresponding RWG edge elements in order to program the condition of zero electric current in the feeds.

6.7. DIRECTIVITY CONTROL

In our discussion of basic antennas in the previous chapters, we saw that these antennas can usually be made more directional by making them large relative to a wavelength. This often poses problems, however, because such structures can become difficult to fabricate and maneuver when they are large. Another problem is that the antennas often do not offer as much freedom as we would like in shaping the exact characteristics of the radiation patterns, such as their directivity and side-lobe characteristics.

An attractive way around these limitations is to construct arrays of small, simple elements, such as dipoles or monopoles. By positioning and feeding each element appropriately, one can attain large directivities, even when the radiation pattern of each element (alone) is relatively poor. Also one can change the direction of maximum radiation by changing the phases of the feed voltages (*electronic beam steering*). Electronic beam steering is often better than mechanical steering, since it is usually faster and does not demand heavy positioning equipment. In this and in the following sections, we will study the directivity and gain control of simple antenna arrays. The beam-steering mechanism (scanning array) will be considered later in this chapter.

There are three major factors that must be considered to control the directivity: (1) the geometrical factor, (2) the phase factor, and (3) the number of array elements. Accordingly there are two basic array types that can be controlled either by element spacing or by relative phases of the fed voltages. These arrays are called *broadside* and *end-fire* arrays, respectively. Broadside arrays generate their maximum radiation perpendicular to the array axis. On the contrary, an end-fire array directs its main lobe along the array axis. Figure 6.6 provides the corresponding schematic.

In terms of the azimuthal angle, ϕ, the *broadside directions* are the $\phi = 90°$ and $270°$ directions in Fig. 6.6. The *end-fire directions* are the $\phi = 0°$ and $180°$ directions, respectively. While the broadside array can be controlled by the array spacing only, keeping the feed voltage the same for any array element, the end-fire array requires voltage phase adjustment. Below we will analyze both array types, with the help of the Matlab scripts of the present chapter.

6.8. BROADSIDE ARRAY

After the code sequence `rwg1.m` - `rwg4.m` is complete, the array radiation pattern and the gain can be calculated using the script `efield3.m`. This script

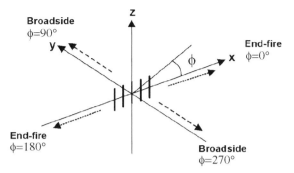

Figure 6.6. Braoadside and end-fire radiation directions for a linear array whose axis is the x-axis.

(run it after efield2.m) outputs only one radiation pattern in the most interesting xy-plane—the H-plane for the array of linear dipoles. The script outputs the array directivity pattern in dB as well as the array gain *for this pattern* (maximum directivity in dB in the xy-plane). Figure 6.7 shows the script output for a two-dipole linear array with the element spacing $d = 0.2\lambda, 0.5\lambda, 1.0\lambda$, and 1.5λ (the offset in dB is set to zero). The dipole length always equals half-wavelength (strip1.mat; frequency is 75 MHz). The x-axis (the antenna axis) corresponds to the zero azimuthal angle.

First we see in Fig. 6.7 that the element spacing greatly affects both the shape of the radiation pattern and the gain. When the separation distance between two dipoles is small (less than a half wavelength), the array does not radiate effectively. For the array with $d = 0.2\lambda$, the gain is 2.8 dB, which is only 0.6 dB greater than the gain of the single half-wavelength dipole (Section 4.9 of Chapter 4). Note that the logarithmic gain of 2.8 dB corresponds to the linear gain of $10^{2.8/10} = 1.91$.

If the spacing between two dipoles becomes greater than one half-wavelength (Fig. 6.7c, d), then the lobes in the broadside direction become narrower. However, the grating lobes (or the *side lobes*) appear, and the overall gain drops down. The best choice is thus $d = \lambda/2$ (Fig. 6.7b), which yields the highest gain of 5.9 dB. Simultaneously, the radiation pattern does not have side lobes. The general rule is therefore that the maximum spacing between the elements should be less than one wavelength [4, pp. 262–266] or even less than one-half wavelength [8, p. 611] to avoid any grating lobes.

A very important figure-of-merit of an array (and any other antenna) is the *half-power beamwidth*, HP. It is the angular separation of two points in Fig. 6.7, where the radiation intensity is equal to one-half of its maximum value. Hence

$$HP = |\varphi_{\text{left}}^{\text{HP}} - \varphi_{\text{right}}^{\text{HP}}| \qquad (6.7)$$

On the logarithmic scale, $10 \times \log_{10}(1/2) \approx -3\,\text{dB}$. Therefore two angles $\varphi_{\text{left}}^{\text{HP}}$, $\varphi_{\text{right}}^{\text{HP}}$ correspond to the—3 dB drop down with regard to the pattern maximum.

126 ANTENNA ARRAYS: THE PARAMETER SWEEP

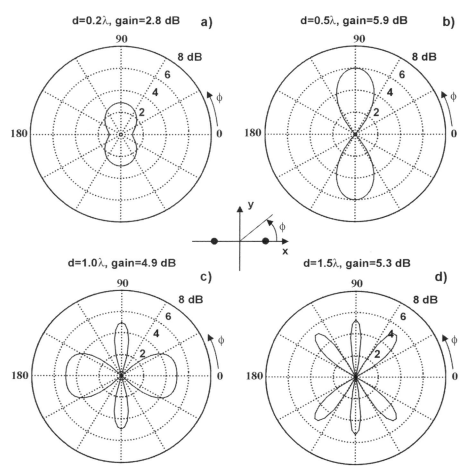

Figure 6.7. Broadside array: Directivity pattern (xy-plane) of the radiated array field (efield3.m). The broadside array of two-half-wavelength dipoles is considered at d = 0.2λ, 0.5λ, 1.0λ, and 1.5λ.

For example, the half-power beamwidth in Fig. 6.7*b* is approximately 60°. In Fig. 6.7*a* the beamwidth is approximately 180° (omnidirectional antenna).

The next optimization parameter is the number of array elements. Figure 6.8 shows the far-field intensity for two arrays with 2 and 5 half-wavelength dipoles, respectively. These arrays are created using the script multilinear.m with a variable number of array elements (N=2 and N=5). The calculations are done in the script efield2.m. The dipole length equals a half-wavelength (frequency is 75 MHz) and the separation distance is $d = \lambda/2$ = 2 m. We see that the array gain increases from approximately 6 to 10 dB if the number of elements increases from 2 to 5.

BROADSIDE ARRAY 127

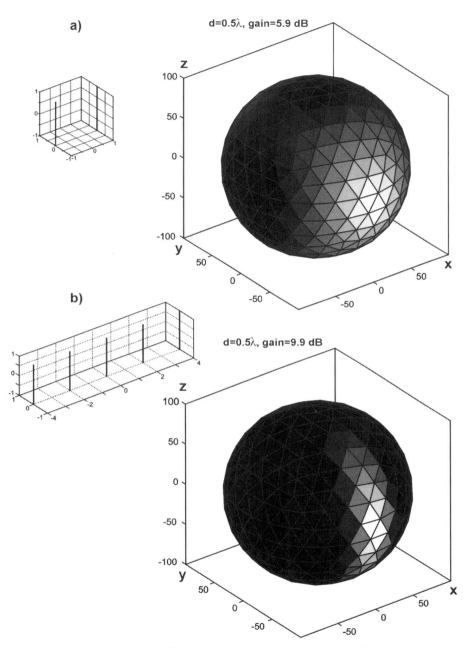

Figure 6.8. Broadside array: radiation intensity over a sphere surface (output of the script efield2.m. (a) Array of two half-wavelength dipoles; (b) array of five half-wavelength dipoles.

128 ANTENNA ARRAYS: THE PARAMETER SWEEP

6.9. END-FIRE ARRAY

An end-fire array directs its main lobe along the array axis (the x-axis in our case). Unlike broadside arrays, in which the phase, δ, of the feed voltage is always the same (zero) for every antenna element, there are a number of phase shifts here that result in end-fire radiation patterns. The phase control is done in the script rwg4.m. Each array element, already sorted in ascending order according to its position on the x-axis, acquires an *incremental* (*progressive*) phase shift according to the following loop:

```
phase=-2*pi/3;
for n=1:N
    nn=Index(n);
    V(nn)= V(nn)*exp(j*phase*(n-1));
end
```

For example, for an array of N elements, this code predicts the feed voltage of the first element as $1 \times \cos(\omega t)$ V, the feed voltage of the second element as $1 \times \cos(\omega t - 120°)$ V, the feed voltage of the third element as $1 \times \cos(\omega t - 240°)$ V, and so on.

How must the phase shift be chosen in order to ensure the radiation in the end-fire direction? Figure 6.9 shows a numerical experiment for a linear array of two dipoles (strip1.mat) involving four different phase shifts: $\delta = 0$, $\delta = -60°$, $\delta = -120°$, and $\delta = -180°$ between two elements. The element spacing is fixed as $d = \lambda/3 = 1.333$ m. The dipole length equals half-wavelength (frequency is 75 MHz). The radiation patterns in the xy-plane are shown. The x-axis (antenna axis) corresponds to an azimuthal angle of zero.

We see from Fig. 6.9 that the nonzero phase shift drastically changes the radiation pattern of the array. When $\delta = 0$, the lobes are directed broadside to the array axis. This case is referred to as a broadside case [1, p. 262], as we saw in the previous section. As δ increases to 180°, the lobes bend toward the end-fire direction $\phi = 0°$. This is the second special case, called the end-fire case [1, p. 264]. The further increase in the phase leads to the same results. The shift in the radiation pattern shown in the figure is a simple example of electronic beam steering.

The most inviting choice in Fig. 6.9 is, however, $\delta = -120°$, when the antenna radiates in only one direction. This choice is also supported theoretically [4, pp. 264–266]. That is to say, signals from the first and the second antenna element arrive at a point in the end-fire direction $\phi = 0°$ in phase, if

$$\delta = -kd = -\frac{2\pi}{3} \tag{6.8}$$

Here $k = \omega/c$ is the wavenumber, and $d = \lambda/3 = 1.333$ m is the separation distance between antenna elements.

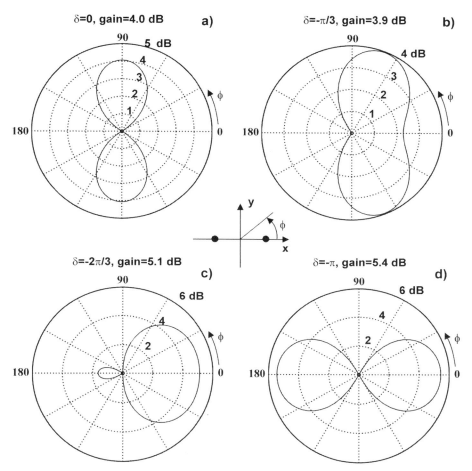

Figure 6.9. (a) Broadside; (b–d) end-fire arrays: Directivity patterns (xy-plane) of the radiated array field (efield3.m). The array of the two half-wavelength dipoles is considered at $\delta = 0$, $\delta = -60°$, $\delta = -120°$, and $\delta = -180°$. The element spacing is d = $\lambda/3$ = 1.333 m.

The term "in phase" means that these signals are added to each other in the end-fire direction $\phi = 0°$. At the same time they are canceled in the opposite direction $\phi = 180°$. This is why we see almost no backward radiation in Fig. 6.9c. Indeed, the interaction between two dipoles modifies the radiation pattern so that we still see weak radiation in the direction $\phi = 180°$.

By analogy with the broadside array, one could expect that if the number of array elements increases, a shaped end-fire beam pattern would be obtained, a pattern better than the end-fire pattern for two dipoles shown in Fig. 6.9c. This is generally true but the array spacing has to be adjusted properly. The end-fire (phased) array is a considerably more complicated radiating system than the broadside array.

130 ANTENNA ARRAYS: THE PARAMETER SWEEP

The broadside array actually is a special case of the phased array where the incremental or progressive phase shift, δ, is equal to zero. The general rule for end-fire arrays is that the maximum spacing between the elements should be less than one quarter-wavelength, namely $d \leq \lambda/4$ [8, p. 613], in order to avoid high lobes in the other directions. Note that this condition is stronger than the corresponding condition $d \leq \lambda/2$ for broadside arrays (see the preceding section).

The increase in directivity of the end-fire multiple-element array is usually more modest than what occurs for broadside arrays [8, p. 613]. The radiated power of end-fire arrays is also smaller.

6.10. PATTERN MULTIPLICATION THEOREM

The basis of the array theory is the pattern multiplication theorem. This theorem states that the combined pattern of N identical array elements can be expressed as the element pattern times an *array factor*,

$$\Lambda(\theta, \phi) \tag{6.9}$$

Here θ, ϕ are the polar and azimuthal angles shown in Fig. 6.10. We will derive the pattern multiplication theorem in the azimuthal plane, namely at $\theta = 90°$, keeping in mind that the same derivation holds in the elevation plane (e.g., see [4, pp. 257–259]). Let us assume that a signal from the first array element has phase zero at an observation point located very far from the array. The corresponding phase factor will be e^{j0}. The signal from the second element will have the relative phase

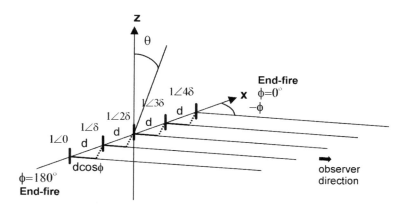

Figure 6.10. A linear five-element array of dipoles. The element spacing is d, the phase shift between adjacent elements is δ, and the observer angle with respect to the array axis (the x-axis) is $-\phi$.

PATTERN MULTIPLICATION THEOREM

$$\psi = kd\cos\phi + \delta \qquad (6.10)$$

at the observation point. Equation (6.10) gives the sum of two components. One of them is the incremental phase shift, δ, in the array element feed. The other is the phase shift arising from the different propagation distances to the observation point. The propagation distance for the second element is smaller by $d\cos\phi$ (Fig. 6.10). The corresponding phase shift is that difference multiplied by the wavenumber, k. Therefore the phase factor of the second element will be $e^{j\psi}$. Analogously the phase factor of the third element is $e^{j2\psi}$, and so it goes. The field at the observation point is the sum of all the element contributions, which yields

$$\Lambda(\psi) = e^{j0} + e^{j\psi} + e^{j2\psi} + \ldots + e^{j(N-1)\psi} \qquad (6.11)$$

where N is the number of antenna elements. To simplify the preceding expression for $\Lambda(\psi)$, let us multiply both sides by $e^{j\psi}$, obtaining

$$e^{j\psi}\Lambda(\psi) = e^{j\psi} + e^{j2\psi} + \ldots + e^{jN\psi} \qquad (6.12)$$

Subtracting this expression from Eq. (6.11) yields

$$(1 - e^{j\psi})\Lambda(\psi) = 1 - e^{jN\psi} \qquad (6.13)$$

Solving for the array factor, we obtain

$$\Lambda(\psi) = \frac{1 - e^{jN\psi}}{1 - e^{j\psi}} = \frac{e^{jN\psi/2}}{e^{j\psi/2}} \frac{e^{-jN\psi/2} - e^{jN\psi/2}}{e^{-j\psi/2} - e^{j\psi/2}} = e^{j\psi[(N-1)/2]} \frac{\sin(N\psi/2)}{\sin(\psi/2)} \qquad (6.14)$$

Finally, since we are usually concerned only with the magnitude of the far-zone radiation pattern, we can drop the exponential phase term, yielding

$$\Lambda(\psi) = \left|\frac{\sin(N\psi/2)}{\sin(\psi/2)}\right| \qquad (6.15)$$

Note that $\Lambda(\psi)$ attains a maximum value of $\Lambda_{max} = N$ when $\psi = 0$. The array factor $\Lambda(\psi)$ of a linear array is a function of the number of elements, their spacing, and the phase difference between each element. Even if the number of elements is fixed, there are a great variety of different array factors that can be obtained, depending on the element spacing d and differential phase shift δ.

Once the array factor is calculated using Eq. (6.15), the beam pattern is obtained as a product of $\Lambda(\psi)$ and the beam pattern of the individual array element. For the half-wavelength dipole considered in this chapter, the beam pattern in the xy-plane is a circle whose radius is 2.15 dB (omnidirectional antenna; cf. Chapter 4). Therefore, the theoretical beam pattern in the xy-plane is the following:

$$D = 10\log_{10}(\Lambda(\psi)) + 2.15\,\text{dB} \qquad (6.16)$$

It must be emphasized that Eqs. (6.15) and (6.16) assume that the array elements are *uncoupled*, meaning that a current in one element does not excite appreciable currents on other elements. This is the most serious limitation of the pattern multiplication theorem, restricting its use to arrays with large element spacing. Equation (6.16) is programmed in the script efield4.m in the Matlab directory of the present chapter.

6.11. COMPARISON OF THEORY AND SIMULATION

In order to compare theory with the numerical simulation, we should minimize coupling between array elements. Ideally an array with infinitely large element spacing should be used. However, such an array does not exist in practice. Therefore the following example is considered. An array of four half-wavelength dipoles at 75 MHz (each dipole is 2 m long) has the element spacing $d = 10$ m and the incremental phase shift $\delta = -180°$ (in order to satisfy the end-fire array condition Eq. (6.8)). Figure 6.11a shows the corresponding radiation pattern in the xy-plane. The solid curve is the numerical simulation, whereas the dashed curve is generated using Eq. (6.16). We see that the agreement between theory and computations is generally very well, including faithful reproduction of the relatively narrow side lobes in the broadside direction. Note that the pattern is calculated at 1000 m.

The pattern multiplication theorem looses its accuracy when the element spacing becomes smaller and the coupling between different array elements becomes important. The numerical approach discussed in this chapter is more general and allows for arbitrary element spacing, including dense arrays. As an example, Fig. 6.11b presents the beam pattern of an array of four half-wavelength dipoles (each 2 m long) separated by $d = \lambda/3 = 1.333$ m at 75 MHz. The phase shift given by Eq. (6.8) is $\delta = -120°$. The solid line is the numerical simulation, whereas the dashed line is generated using Eq. (6.16). One can see that the array gain remains nearly the same in both the cases. However, Eq. (6.16) predicts a higher beamwidth and some side lobes. These results are not confirmed by the numerical simulation.

As a further example, we consider the same array of four dipoles when the element spacing reduces to $d = \lambda/4 = 1$ m at 75 MHz. The corresponding phase shift given by Eq. (6.16) is $\delta = -90°$. The disagreement between the theory data (dashed line) and simulation (solid line) in Fig. 6.11c becomes even worse. It also involves a considerable discrepancy in the predicted array gain.

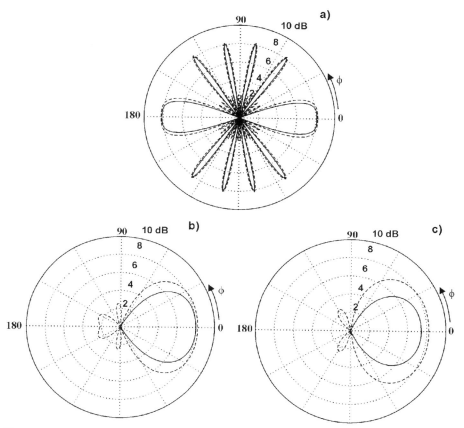

Figure 6.11. (a) Beam pattern for a four-element end-fire linear dipole array at 75 MHz when d = 10 m (solid line—numerical simulation; dashed line—theory prediction); (b) beam pattern for a four-element end-fire linear dipole array at 75 MHz when d = 1.333 m (solid line—numerical simulation; dashed line—theory prediction); (c) the same patten as (b) but when d = 1 m and $\delta = -90°$. The patterns are calculated at 1000 m.

6.12. OPTIMIZATION OF END-FIRE ARRAY: THE PHASE LOOP

A typical optimization problem for the phased array implies high directivity, narrow beamwidth, and low side lobes. A large number of theoretical approaches have been developed to solve this problem [9]. Among them is the Hansen-Woodyard formula for the phase shift of closely spaced array elements [4, pp. 271–276]. In this approach only the phase is optimized, whereas other array parameters are fixed. The Hansen-Woodyard formula takes the form

$$\delta = -\left(kd + \frac{\pi}{N}\right) \qquad (6.17)$$

134 ANTENNA ARRAYS: THE PARAMETER SWEEP

where N is the number of array elements. Equation (6.17) assumes that the number of array elements is large, namely $N \gg 1$. This formula gives the best results if the spacing between array elements is approximately $\lambda/4$, that is,

$$d = \frac{\lambda}{4} \qquad (6.18)$$

More sophisticated (binomial [4] and Taylor [10]) distributions also vary the amplitude of the feed voltage along the array axis. The general idea for the nonuniform feed voltage is a tapered amplitude profile, with almost zero voltage at the array ends [11]. A nonuniform element spacing can be also used to optimize the beam pattern [12]. A number of analytical results for planar arrays are collected in [13] (see also the corresponding reference list) and in [14, ch. 9].

The numerical simulation gives the most accurate optimization results for any type of the array. However, this poses a problem, which is connected to a large number of beam patterns that have to be generated and observed for the optimization purposes. In this section we discuss the generation and graphical representation of beam patterns corresponding to different phase shifts (the phase loop). The element spacing, d, of the array is assumed to be fixed. The idea of the phase loop can be extended to model a nonuniform amplitude distribution along the array as well.

The organization of the Matlab loop has already been outlined in Section 6.5. There a loop with respect to different element spacing of an array was created. The corresponding loop codes were saved in subdirectory `loop1` of the Matlab directory of Chapter 6. For the phase loop, we reserve the subdirectory `loop2`. The loop variable is now the incremental phase shift, `phase(k)`.

As in Section 6.5, the corresponding Matlab scripts are first converted into Matlab functions by adding a `function` line at the top of them. In particular, the function `multilinear` has now two arguments:

`multilinear(d,N);`

The first argument is the element spacing, whereas the second argument denotes the number of elements. A wrapper script (`loop.m`) yields the anticipated phase loop in the following form:

```
K=20; Step=pi/K; %rad
for k=1:K+1
    phase(k) =-(k-1)*Step;
    Power1   =rwg4(phase(k));
    [PolarXY, GAIN, phi] =efield3;
    Radius   =max(PolarXY,0);
    x1(k,:)  =Radius.*cos(phi);
    y1(k,:)  =Radius.*sin(phi);
    z1(k,:)  =-phase(k)*ones(length(x1),1)';
end
```

The loop calculates the beam pattern for each phase value in the range from 0 to $-\pi$. These beam patterns are originally plane curves in the xy-plane. If we assign a coordinate $z = -\text{phase}(k)$ to each pattern, we obtain K spatial curves. Taken together, these curves form a surface in 3D. The surface is visualized and observed (using mouse rotation) using the following command line:

```
surf(x1,y1,z1); colormap gray; rotate3d;
```

Thus a beam pattern at a given phase(k) is obtained as a cross section of the output surface by a plane perpendicular to the z-axis.

6.13. HANSEN-WOODYARD MODEL

Figure 6.12 shows the output of the script `loop.m` (phase loop in subdirectory `loop2`) for the array of two half-wavelength dipoles (`strip1.mat`) at 75 MHz separated by $d = \lambda/4 = 1$ m. To obtain Fig. 6.12, run this script and do not change anything in the code.

The phase is now measured in radians. The beam pattern at $\delta = 0$ is the broadside array; the beam pattern at $\delta = -\pi$ is the end-fire array with two opposite side lobes. The solid curve indicates the radiation pattern at $\delta = -\pi/2$, obtained according to Eq. (6.8). Note that the Hansen-Woodyard formula (6.17) predicts a rather incorrect phase value $\delta = -\pi$ in that case. The reason for this is the insufficient number of array elements, N. Figure 6.12 indicates that $\delta = -\pi/2$ is almost the best choice for the phase shift when the condition of no opposite side lobe is implied. A slightly higher (in magnitude) δ will increase the gain in the end-fire direction $\phi = 0°$ but simultaneously will create a minor side lobe in the opposite direction.

The next example is the same array but the number of dipoles is increased from 2 to 4. Figure 6.13 shows the corresponding output of the script `loop.m`. Forty sampling points (K=40) in the phase domain are used instead of 20 points in the previous example. This is necessary because the phase discretization and the discretization of the beam pattern have to be refined in order to see the details of the pattern transformation. The solid curve indicates the radiation pattern at $\delta = -\pi/2$, obtained according to Eq. (6.8); the dashed line is the pattern at $\delta = -3\pi/4$ (Hansen-Woodyard formula).

Similar to the array of two dipoles, the Hansen-Woodyard formula predicts too high magnitude of the phase shift. The optimum phase shift lies somewhere between $\delta = -\pi/2$ (Eq. (6.8)) and $\delta = -3\pi/4$ (Hansen-Woodyard formula).

In the final example (Fig. 6.14), the number of dipoles is increased to eight. Figure 6.14 shows the corresponding output of the script `loop.m`. Unlike Figs. 6.12 and 6.13, different rotation angles are used in order for us to better observe the beam pattern transformation. The solid curve is again the radiation pattern at $\delta = -\pi/2$. The dashed curve corresponds to $\delta = -5\pi/8$. This value is obtained using the Hansen-Woodyard formula. We can see from Fig. 6.14

136 ANTENNA ARRAYS: THE PARAMETER SWEEP

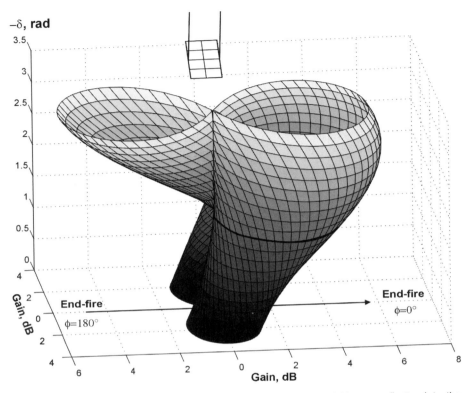

Figure 6.12. Transformation of the radiation pattern from a broadside array (bottom) to the end-fire array (top) as a function of the phase shift. The array of two dipoles separated by d = λ/4 is considered. The solid curve indicates the radiation pattern at $\delta = -\pi/2$.

that the Hansen-Woodyard formula finally gives the result much better than Eq. (6.8). In other words, we have "enough" array elements to be sure that Eq. (6.17) is working properly. Minor side lobes still are expected at $\delta = -5\pi/8$. The optimum phase shift could have a smaller magnitude.

It is interesting to test the radiation pattern for $\delta = -5\pi/8$. The phase loop is not appropriate for the generation of single patterns. Therefore the core code in the root directory Matlab should be used. Figure 6.15a shows the beam pattern of the eight-element end-fire array at $\delta = -5\pi/8$. The array gain is 11.3 dB. We also see a number of side lobes whose strength reaches 5 dB.

For the purposes of comparison, Fig. 6.15b shows a similar result obtained for the eight-element array if we change the separation distance to $d = \lambda/3$. The phase shift is chosen in the form $\delta = -0.72\pi$. The radiation pattern surprisingly has a higher gain and smaller side lobes. This is despite the fact that the value $d = \lambda/3$ does not actually belong to the "design" range $d \leq \lambda/4$ [4,8].

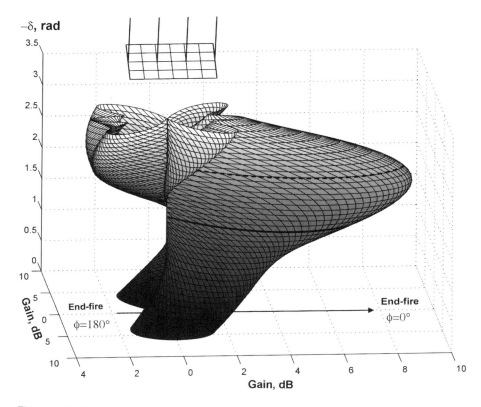

Figure 6.13. Transformation of the radiation pattern from a broadside array (bottom) to the end-fire array (top) as a function of the phase shift. The array of four dipoles separated by d = $\lambda/4$ is considered. The solid curve indicates the radiation pattern at $\delta = -\pi/2$; the dashed line corresponds to Eq. (6.17).

We see that the phased array design could show unique behavior, especially in the case of mutually coupled array elements.

The phase loop developed in this section allows for arbitrary phase distributions, and not just for the uniform progressive phase shift. Similarly a nonuniform amplitude distribution along the array can be considered such as binomial or Dolph-Tschebyscheff distribution [4, ch. 6].

6.14. POWER MAP OF END-FIRE ARRAY

A clue for understanding the physics of end-fire arrays is the distribution of radiated power for different array elements. Since we consider lossless arrays, the array `FeedPower(1:N)` from the script `rwg4.m` is equivalent to the distribution of radiated power. Figure 6.16 outlines the radiated power for each

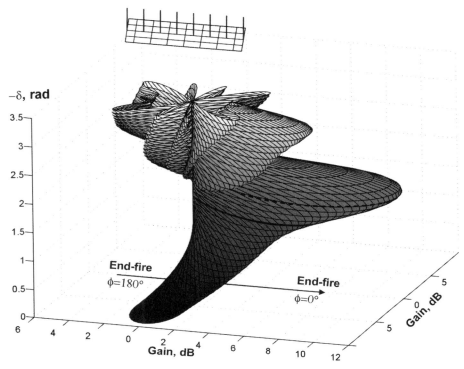

Figure 6.14. Transformation of the radiation pattern from a broadside array (bottom) to the end-fire array (top) as a function of the phase shift. The array of eight dipoles separated by $d = \lambda/4$ is considered. The solid curve indicates the radiation pattern at $\delta = -\pi/2$; the dashed line corresponds to Eq. (6.17).

of the elements of the end-fire array of eight half-wavelength dipoles (strip1.mat) with the spacing $d = \lambda/4$ and $\delta = -\pi/2$. The dashed line is the power radiated by a single dipole antenna. The radiation pattern of that array is outlined by a solid line in Fig. 6.14.

As can be seen in the figure, only the leading array element radiates substantial power. The other array elements radiate very poorly. In other words, they squint the already radiated electromagnetic field in the end-fire direction $\phi = 0°$.

Actually we do not even need the voltage feeds at these almost passive elements and can replace the elements by simple wires. This idea is realized in the Yagi-Uda antenna—a very practical directional radiator (and receiver) in the broad range 3 MHz –3 GHz [4, pp. 513–532]. The Yagi-Uda antenna was widely used as a home TV antenna, so it should be familiar to most readers.

Since the voltage phase control in multiple array feeds is expensive (see the next section) a reasonable question arises of why do we need the end-fire arrays at all. The reason for using end-fires arrays is, nevertheless, an ability of electronic beam steering. An example will be considered in the next section.

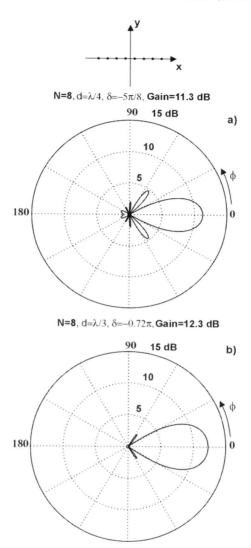

Figure 6.15. Beam pattern for the eight-element end-fire linear array; (a) d = λ/4 and δ = −5π/8; (b) d = λ/3 and δ = −0.72π.

6.15. PHASED (SCANNING) ARRAY

The idea of the scanning array is very simple and elegant [4,9]. The maximum value of the array factor given by Eq. (6.15) is attained when (Section 6.10)

$$\psi = kd\cos\phi + \delta = 0 \tag{6.19}$$

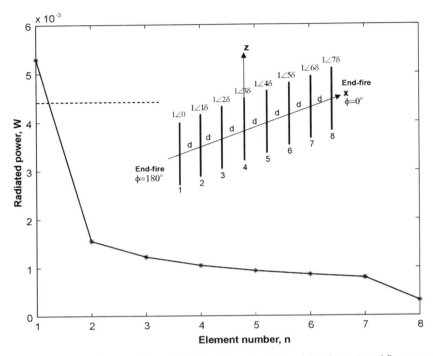

Figure 6.16. Radiated power for each of the elements of the eight-element end-fire array of half-wavelength dipoles with d = λ/4 and δ = −π/2. The dashed line shows the power radiated by a single array element—the half-wavelength dipole.

Here ϕ is the azimuthal angle in the xy-plane and δ is the incremental or progressive phase shift. Let us assume that the maximum radiation of the array is required to be oriented at an angle ϕ_0. To accomplish this, the phase excitation, δ, between the elements must be adjusted so that

$$kd\cos\phi_0 + \delta = 0 \quad \text{or} \quad \delta = -kd\cos\phi_0 \tag{6.20}$$

Thus, by controlling the progressive or incremental phase difference between the elements, the maximum radiation can be directed at any desired angle to form a scanning array. This is the basic principle of electronic scanning phased array operation. Since, in the phased array technology, the scanning must be continuous, the system should be capable of continuously varying the progressive phase between the elements. In practice, this is done electronically by the use of ferrite or diode phase shifters. For ferrite phase shifters, the phase shift is controlled by the magnetic field within the ferrite, which in turn is controlled by the amount of current flowing through the wires wrapped around the phase shifter.

For a diode phase shifter using balanced, hybrid coupled variable capacitors (varactors), the actual phase shift is controlled either by varying the bias dc voltage or by a digital command using a DAC (digital-to-analog) converter [4, ch. 6].

To demonstrate the principle of scanning, we use the code sequence in the root directory Matlab. An eight-element array of half-wavelength dipoles (strip1.mat) at $d = \lambda/4$ and with a starting phase shift $\delta = -\pi/2$ is considered. Script efield2.m outputs the radiation intensity distribution over a large sphere and allows us to visualize the scanning directions. Note that we will be using the finer structure sphere1.mat with 8000 boundary elements in the script efield2.m. Both sphere structures (sphere.mat and sphere1.mat) are saved in the Matlab root directory of Chapter 6. Figure 6.17 shows four intensity distributions corresponding to $\phi_0 = 0°, 30°, 45°,$ and $60°$. This gives, according to Eq. (6.20), $\delta = -0.500\pi, -0.433\pi, -0.354\pi,$ and -0.250π. Rotate the sphere for the best observation angle.

In Fig. 6.17 we can observe how the maximum radiation (and reception) directions of the array track the angle $\pm\phi_0$ in the azimuthal plane by way of a variable progressive phase shift. Indeed, a commercial phased array would provide better performance and narrower beamwidth, but that assumes a more sophisticated design.

6.16. ARRAY OF BOWTIES OVER A GROUND PLANE

The last example of the present chapter is rather practical. We consider an array of 8×8 small bowtie antenna elements over a ground plane. This design relies upon one of the prototypes developed by Seavey Engineering Associates, Inc., Massachusetts. The array structure is shown in Fig. 6.18.

The mesh for the antenna array is created using the array generator multibowtie.m in subdirectory mesh. This scripts imports the bowtie mesh (bowtie.mat)[2] and clones it into the array of 8×8 mutually perpendicular array elements. Further it creates a ground plane (plate.mat)[3] and attaches it underneath to the array of bowties. The interested reader can observe the structure shown Fig. 6.18a if he or she runs the script multibowtie.m and does not change anything in the original code. The script allows us to vary array dimensions, number of array elements, and relative spacing.

The following dimensions are chosen for the present simulation: the size of the ground plane is 6×6 m; the size of the bowtie array element is 0.45×0.5 cm (flare angle is $90°$); the separation distance between the array and the ground plane is 0.5 m. The operation frequency is 150 MHz.

Each bowtie is divided into 30 subtriangles; the plate has 512 subtriangles. The total number of RWG edge elements of the structure is 2976 (the number

---

[2] For the bowtie mesh generator, see Chapter 7.
[3] For the plate/strip mesh generator, see Chapter 4.

142 ANTENNA ARRAYS: THE PARAMETER SWEEP

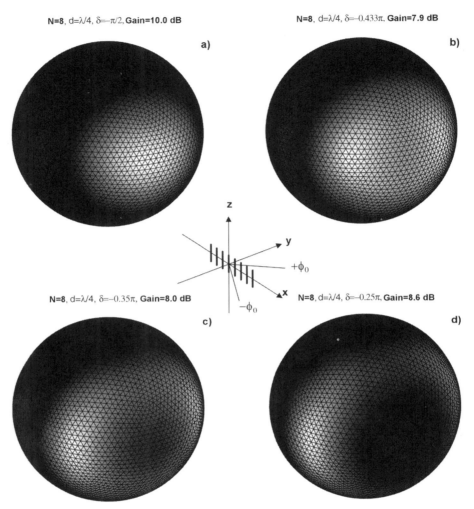

Figure 6.17. Radiation intensity distribution for a scanning array of eight half-wavelength dipoles separated by d = $\lambda/4$ at different values of the progressive phase shift, δ.

of triangles is 2432). Each bowtie is a center-fed (dipole-type) antenna, with a feeding edge located exactly in the middle junction (Fig. 6.18c). The calculations are performed in the root directory Matlab after all the antenna parameters are specified. The phase angles are equal to zero.

The code for that array requires relatively large CPU processor run times. In particular, the script rwg3.m is executed in 4.8 minutes; the script rwg4.m—in 2.2 minutes. The whole code sequence takes about 9 minutes. These data are for a computer with Pentium IV 1.7 GHz processor and 1 Gbyte RAM (Windows ME).

ARRAY OF BOWTIES OVER A GROUND PLANE   143

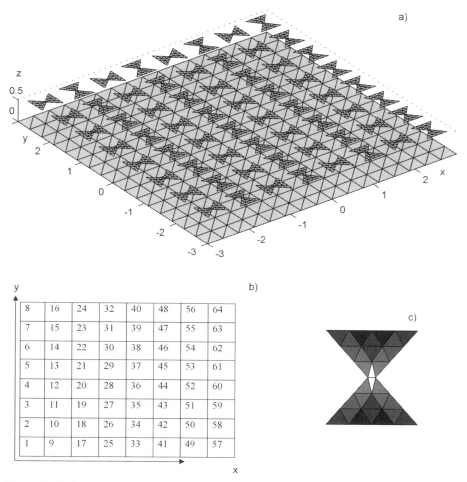

Figure 6.18. (a) Array of 8 × 8 bowties over a ground plane; (b) map of the element numbers; (c) mesh for a single bowtie and a typical surface current distribution.

Figure 6.19 shows the surface current distribution obtained using the script rwg5.m at 150 MHz (the top view). This map is simultaneously the map of the radiated power, since power is current times voltage in the feed, and the voltage is always equal to 1 V. It is seen that almost all array elements create substantial surface currents (radiate effectively). However, there are some "dark" elements that do not radiate substantial power. The situation, however, improves when frequency reduces to 120–130 MHz. The analysis of the corresponding current and power (array FeedPower(1:N)) distributions at 120 MHz show nearly uniform power distribution across the entire array lattice.

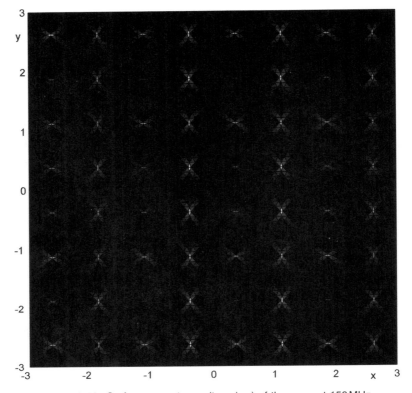

Figure 6.19. Surface current map (top view) of the array at 150 MHz.

Table 6.1 gives the input impedance (terminal impedance) values of the array elements at 150 MHz. The element mapping for this table corresponds to Fig. 6.18b. Obviously the elements with larger impedances have smaller current magnitudes, since $J^n = 1\,\text{V}/Z_n^{in}$.

Figure 6.20 shows the 3D radiation pattern of the array at 150 MHz (efield2.m) and the corresponding directivity pattern in the xz-plane (a slightly modified efield3.m to account for the xz-plane instead of the xy-plane). Although the main beam has a significant gain and a small half-power beamwidth, side lobes are observed on the level of −14 dB.

6.17. ON THE SIZE OF THE IMPEDANCE MATRIX

The preceding section posed a problem connected to the maximum size of the impedance matrix allowed by Matlab. An attempt to increase the mesh size for the single bowtie to 48 triangles led to a structure totaling 3584 triangles and 4512 RWG edge elements. Although the impedance matrix can

Table 6.1. Terminal Impedance at 150 MHz, ×100 Ω

0.19 −0.36j	0.34 −0.33j	0.27 −0.43j	0.24 −0.28j	0.24 −0.41j	0.24 −0.28j	0.26 −0.41j	0.35 −0.32j
0.76 −0.71j	0.44 −0.24j	−0.30 −1.92j	0.29 −0.22j	0.80 −2.40j	0.29 −0.21j	0.88 −0.85j	0.42 −0.24j
0.29 −0.48j	0.47 −0.34j	0.24 −0.47j	0.32 −0.25j	0.26 −0.44j	0.31 −0.25j	0.22 −0.46j	0.50 −0.30j
0.27 −0.59j	0.40 −0.29j	0.39 −0.83j	0.27 −0.23j	0.41 −0.87j	0.28 −0.23j	0.55 −0.71j	0.44 −0.30j
0.53 −0.56j	0.44 −0.31j	0.29 −1.03j	0.27 −0.23j	0.44 −0.87j	0.28 −0.22j	0.45 −0.66j	0.38 −0.31j
0.23 −0.43j	0.47 −0.32j	0.24 −0.45j	0.32 −0.25j	0.26 −0.44j	0.31 −0.26j	0.22 −0.45j	0.54 −0.29j
0.34 −0.99j	0.40 −0.22j	0.80 −2.06j	0.30 −0.22j	0.39 −2.23j	0.29 −0.22j	1.15 −1.02j	0.42 −0.22j
0.31 −0.36j	0.34 −0.36j	0.22 −0.44j	0.24 −0.27j	0.24 −0.41j	0.26 −0.27j	0.26 −0.41j	0.30 −0.34j

still be created and saved on the hard drive, the matrix equation becomes impossible to solve directly (Gaussian elimination). An Out of memory warning message appears that cannot be handled using the standard Matlab help recommendations such as increasing virtual memory from the Windows Control Panel.

One option that has worked for many users in the past who were experiencing Out of memory errors is to disable the Java Virtual Machine that starts up with MATLAB 6.0.[4] The reader can disable Java in MATLAB R12 by starting MATLAB with the -nojvm option. The easiest way to do this is to change the target in the shortcut properties to:

$MATLAB\bin\matlab.exe -nojvm

Testing of the present example has, however, shown that the disabled Java still does not allow one to solve the MoM equation with the complex impedance matrix of 4512 × 4512 in size.

This and other examples indicate that the realistic maximum size of the impedance matrix is 4000 × 4000. This estimate is for a computer with a Pentium IV 1.7 GHz processor and 1 Gbyte RAM (Windows ME).

Indeed, when an iterative solver is implemented instead of the direct solution, the maximum matrix size can be increased. Matlab has a wide range of iterative solvers including conjugate gradient and GMRES methods. The iterative solvers, however, work poorly for complex structures like the metal

---

[4] Reference from Mr. Kevin Shea, the MathWorks Helpdesk.

146 ANTENNA ARRAYS: THE PARAMETER SWEEP

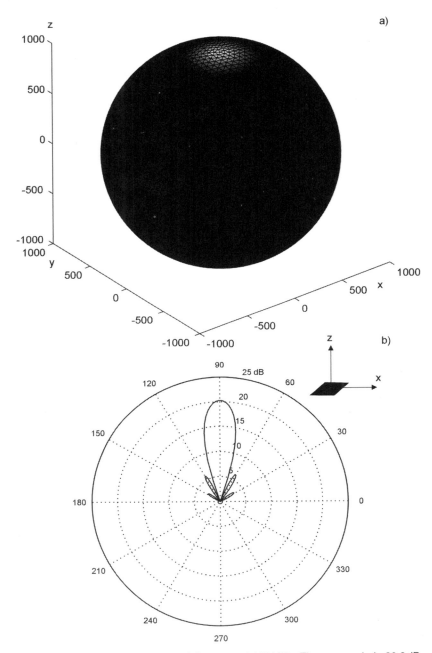

Figure 6.20. 3D and 2D patterns of the array at 150 MHz. The array gain is 20.2 dB.

arrays and metal meshes. The interested reader could test the performance of these solvers by replacing the corresponding line in the script rwg4.m of this chapter.

6.18. CONCLUSIONS

In this chapter we mostly concentrated on the linear arrays of dipoles. While only broadside and end-fire arrays were considered, investigation of circular arrays and the arrays of monopoles can be done in exactly the same way. In the case of a monopole phased (end-fire) array, the script rwg4.m would be slightly changed to account for the same phase shift of two feeding RWG elements.

The present algorithm allows for arbitrarily spaced arrays, not necessary linear or planar. In that case the Matlab array Feed, which describes the positions of the feeding edges, should be modified properly. In particular, the log-periodic array can be created using a straightforward modification of the script multilinear.m. By changing the mesh for the array element, we can create an array of loops, bowties, and the like. For an array with passive elements, like the Yagi-Uda array, some of the antenna feeds should be disabled.

One more important loop that was omitted in this chapter is the frequency loop. Chapter 7 will introduce the frequency loop, but mainly for certain single antenna configurations. The interested reader could change the code sequence of Chapter 7 to make it applicable to arrays. To do so, the array Feed should be included in the mesh files mesh1.m and mesh2.m of the next chapter. Also the array Index in script rwg3.m of Chapter 7 should be defined as described in the script rwg4.m of the present chapter.

REFERENCES

1. J. D. Kraus. *Antennas*. McGraw Hill, New York, 1950.
2. E. Roubine, J. C. Bolomey, S. Drabowitch and C. Ancona. *Antennas*, Vols. 1, 2. Hemisphere Publishing Corporation, Washington, 1988.
3. J. D. Tillman Jr. *The Theory and Design of Circular Antenna Arrays*. University of Tennessee Engineering Experiment Station, 1966.
4. C. A. Balanis. *Antenna Theory: Analysis and Design*, 2nd ed. Wiley, New York, 1997.
5. W. L. Stutzman and G. A. Thiele. *Antenna Theory and Design*. Wiley, New York, 1981.
6. D. M. Pozar. *Microwave Engineering*, 2nd ed. Wiley, New York, 1998.
7. D. M. Pozar. *Microwave and RF Design of Wireless Systems*. Wiley, New York, 2001.
8. K. R. Demarest. *Engineering Electromagnetics*. Prentice Hall, Upper Saddle River, NJ, 1998.

9. R. C. Hansen. *Phased Array Antennas*. Wiley, New York, 1998.
10. T. T. Taylor. Design of line-source antennas for narrow beamwidth and low side-lobes. *IRE Trans. Antennas and Propagation*, 3(1): 16–28, 1955.
11. K. L. Virga and M. L. Taylor. Transmit patterns for active linear arrays with peak amplitude and radiated voltage distribution constraints. *IEEE Trans. Antennas and Propagation*, 49(5): 732–739, 2001.
12. N. Herscovici. Low-sidelobe arrays fed by a uniform-distribution feeding network. *IEEE Antennas and Propagation Magazine*, 39(3): 72–74, 1997.
13. M. J. Lee, L. Song, S. Yoon, and S. R. Park. Evaluation of directivity for planar antenna arrays. *IEEE Antennas and Propagation Magazine*, 42(3): 64–67, 2000.
14. D. H. Staelin, A. W. Morgenthaler, and J. A. Kong. *Electromagnetic Waves*. Prentice Hall, Upper Saddle River, NJ, 1998.

PROBLEMS

6.1. Create and plot the antenna structure for the linear array of six half-wavelength dipoles (strip1.mat) at 75 MHz with element spacing $\lambda/4$.

6.2. Create and plot the antenna structure for the circular array of six half-wavelength dipoles (strip1.mat) at 75 MHz equally spaced around the circumference of a circle with the radius $R = 4$ m.

6.3. Create and plot the antenna structure for the linear array of four quarter-wavelength monopoles at 75 MHz with element spacing $\lambda/8$. The plate size is 2 by 2 m.

6.4. Calculate terminal impedances, $Z^{in}_{1,2}$, of the two-element linear array of half-wavelength dipoles (strip1.mat) at 75 MHz with element spacing $\lambda/4$. The feed voltage, $V^{1,2} = 1$ V. Compare the impedance values with the input impedance of the single half-wavelength dipole antenna at the same frequency.

6.5. Solve Problem 6.4 when the element spacing is equal to 10λ.

6.6. Calculate terminal impedances, Z^{in}_{1-5}, of the five-element linear array of half-wavelength dipoles (strip1.mat) at 75 MHz with element spacing $\lambda/4$. The feed voltage, $V^{1,2} = 1$ V. Compare the impedance values with the input impedance of the single half-wavelength dipole antenna at the same frequency.

6.7. Calculate terminal impedances, $Z^{in}_{1,2,3}$, of the three-element linear array of half-wavelength dipoles (strip1.mat) at 75 MHz as a function of the variable element spacing, d. The element spacing varies as $0.5:0.25:20$ m. The feed voltage, $V^{1,2,3} = 1$ V. Plot impedance magnitude for (a) central dipole; and (b) one of the border dipoles.

6.8. For a linear three-element array of half-wavelength dipoles (strip1.mat) at 75 MHz calculate total radiated power as a function of the variable element spacing, d. The element spacing varies as $0.5:0.25:20$ m. The feed voltage, $V^{1,2,3} = 1$ V.

6.9. Design a broadside linear array of four half-wavelength dipoles (strip1.mat) at 75 MHz. Calculate the following parameters of interest:
  a. Total array impedance at a driving point
  b. Array gain
  c. Half-power beamwidth.

Plot the array radiation pattern in the xy-plane and the radiation intensity distribution over a sphere surface with the radius 1000 m.

6.10.* Solve Problem 6.9 when the number of dipoles is increased to 10. Use the structure sphere1.mat instead of sphere.mat in the script efield2.m in order to plot the radiation intensity distribution over a sphere surface with the radius 1000 m.

6.11. A linear end-fire array of two half-wavelength dipoles (strip1.mat) at 75 MHz has element spacing $\lambda/4$. Plot the radiation pattern in the xy-plane and the radiation intensity distribution over the sphere surface when the progressive phase shift is $\delta = 0, -\pi/2, -\pi$.

6.12. Solve Problem 6.11 if the number of dipoles is increased to 7. What is the half-power beamwidth of the main lobe at $\delta = -\pi/2$?

6.13. Give the array factor, $\Lambda(\psi)$, for the linear end-fire array of six half-wavelength dipoles at 75 MHz with the element spacing $\lambda/4$ and $\delta = -\pi/2$. Plot the azimuthal radiation pattern using the pattern multiplication theorem. Compare this pattern with the corresponding result obtained numerically.

6.14. For a linear end-fire array of five half-wavelength dipoles at 75 MHz with the element spacing $\lambda/4$ plot a set of radiation patterns when the progressive phase shift varies in the range $\delta = [-\pi, 0]$. Use the script loop.m from the subdirectory loop2. Separately plot the radiation pattern corresponding to Hansen-Woodyard model.

6.15. Design a linear end-fire array of seven half-wavelength dipoles with the maximum radiation in the end-fire direction $\phi = 0°$. Calculate the array gain, total radiated power, and the power distribution for different array elements.

6.16. Design a linear end-fire array of seven half-wavelength dipoles with the maximum radiation in the direction $\phi = \pm 45°$. Calculate the array gain, total radiated power, and the power distribution for different array elements.

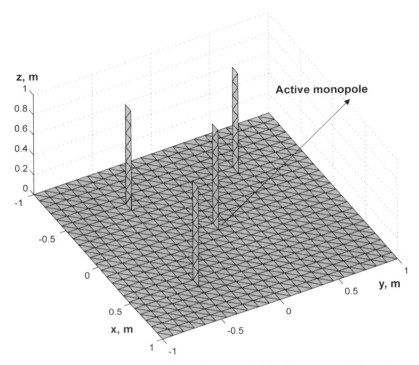

Figure 6.21. Active monopole (center) surrounded by three parasitic elements.

6.17.* For a quarter-wavelength base-driven monopole in the middle of a 2 by 2 m ground plane shown in Fig. 6.21 obtain the radiation intensity distribution over a sphere surface with the radius 1000 m. Compare this result with the intensity distribution of the single monopole, when three parasitic elements are removed.

7

BROADBAND ANTENNAS: THE FREQUENCY SWEEP

7.1. Introduction
7.2. Code Sequence
7.3. Antenna Structures under Study
7.4. Dipole Impedance and Power Resonance
7.5. Dipole Radiated Power, Return Loss, and Gain
7.6. Dipole Comparison with NEC Modeling
7.7. Matlab Mesh for Bowtie Antenna Using Delaunay
7.8. Bowtie Impedance
7.9. Bowtie Radiated Power and Gain
7.10. Bowtie Radiation Intensity Distribution
7.11. Mesh for a Spiral Antenna
7.12. Spiral Antenna's Impedance, Power, and Gain
7.13. Spiral Antenna's Radiation Intensity Distribution
7.14. Multiband Antennas: The Sierpinski Fractal
7.15. Sierpinski Fractal's Impedance, Power, and Gain
7.16. Conclusions
References
Problems

7.1. INTRODUCTION

In the preceding chapters the antenna analysis has been done at a single frequency. In order to obtain antenna parameters at another frequency, the frequency value should be changed in the script rwg3.m and the entire sequence rwg3.m; rwg4.m; rwg5m has to be repeated. Then the far-field results (scripts efield1.m; efield2m; efield3.m) should be updated accordingly.

This procedure is rather inconvenient since the most important parameters of any antenna, both narrowband and broadband, are its characteristics in the frequency domain. Input impedance, radiated power, directivity patterns, and gain should be calculated, as a rule, within a given frequency bandwidth. In this chapter we discuss the organization of the frequency loop and apply the frequency analysis to the simple antenna types, including the dipole antenna (narrowband), the bowtie antenna (broadband), and the spiral antenna (broadband).

In many applications an antenna must operate effectively over a wide range of frequencies. An antenna with wide bandwidth is referred to as a broadband antenna. The term "broadband" is a relative measure of bandwidth and varies with the circumstances. Here we provide a brief definition of *bandwidth* (see also [1, p. 63]). The bandwidth is computed in one of two ways. Let f_U and f_L be the upper and lower frequencies of operation for which satisfactory performance is obtained. The center (or sometimes the design frequency) is denoted by f_C. Then bandwidth as a percent of the center frequency is

$$\frac{f_U - f_L}{f_C} \times 100\% \tag{7.1}$$

This definition holds for narrowband antennas only, such as the dipole or a loop. For example, a 5% bandwidth indicates that the frequency difference of acceptable operation is 5% of the center frequency of the bandwidth. For broadband antennas the bandwidth is defined as a ratio by

$$f_U : f_L \tag{7.2}$$

For example, a 10:1 bandwidth indicates that the upper frequency is 10 times greater than the lower.

We will see that the resonant antennas like dipoles have small bandwidths. For example, the half-wavelength dipoles have typical bandwidth from 8 to 16% [2, p. 260]. On the other hand, antennas that have traveling waves on them rather than standing waves (as in resonant antennas) operate over a significantly wider frequency range. The definition of a broadband antenna is somewhat arbitrary, but we will adopt the following working definition [2, p. 261]: If the impedance and the principal radiation pattern of an antenna do not change significantly over an *octave* ($f_U/f_L = 2$) or more, we will classify the antenna as a *broadband antenna*. For *ultra-wideband* antennas capable of transmitting voltage pulses, the bandwidth of $f_U/f_L \geq 10$ is usually required.

The subject of broadband antennas is well covered in the literature (see [1, chs. 9–11] and [2, ch. 6]). Most conventional broadband antennas include the helical antenna introduced in Chapter 5, the biconical/bowtie antenna, the spiral antenna, the horn antenna, and the log-periodic antenna.

The so-called ultra-wideband (UWB) antennas should not only keep nearly the same impedance and radiation pattern but also have a small phase variation of the transmitted signal over the bandwidth. Otherwise, the pulse form would be distorted because of dispersion of different harmonics. Because of their dispersive characteristics conventional wideband antennas, such as log-periodic arrays, spiral antennas, and ridged horn type antennas perform poorly as high-fidelity antennas for UWB impulse waveforms [3].

Examples of ultra-wideband communication antennas include loaded dipoles [4–6], (loaded) bowties, conical antennas [7–12], and vee configurations [13]. We will study loaded dipoles further in Chapter 9. The bowtie antenna will be considered in the present chapter. A magnetic slot UWB antenna for pulse transmission will be considered in Chapter 8, which includes the corresponding time-domain analysis.

7.2. CODE SEQUENCE

The code sequence of the preceding chapters had two parts. The first part rwg1.m - rwg5.m creates RWG elements (scripts rwg1.m, rwg2.m), performs MoM calculations (scripts rwg3m, rwg4.m) and visualizes surface currents on the antenna surface (rwg5.m). The second part efield1.m - efield3.m found antenna far-field parameters including 3D gain, total radiated power, and directivity patterns.

We can conveniently organize these codes into a frequency loop by combining the frequency dependent MoM codes into a single script rwg3.m. This script now has a loop versus frequency. The bandwidth, sampling rate, and the number of sampling frequencies are specified in that script and can be chosen arbitrarily. The script outputs a binary file current.mat that contains the antenna impedance and surface current distribution for every frequency. Another file containing the identical frequency loop is the script efield2.m. This script outputs the total radiated power and the antenna gain over a surface of a large sphere (sphere.mat) as functions of frequency. If a highly directional antenna is considered, the sphere structure sphere1.mat can be used, with 8000 triangular elements. The script outputs a binary file gainpower.mat that contains the power and gain data for every frequency.

The sequence of operations implies that we run scripts rwg1.m and rwg2.m for a given antenna geometry only once. The loop parameters are then specified in the script rwg3.m. This script and the script efield2.m are executed sequentially (Fig. 7.1).

Frequency-dependent quantities such as antenna impedance, total radiated power (at 1 V feed voltage), and antenna gain are plotted using the script sweeplot.m. Before reading this chapter, you may want to run this script: it will output the precalculated data for the dipole. The rest of the code (Fig. 7.1)

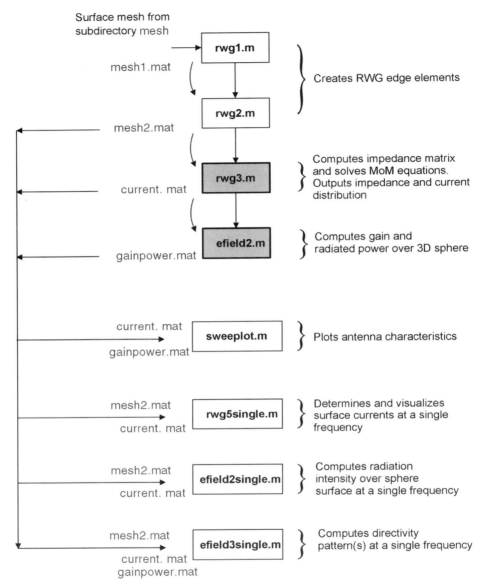

Figure 7.1. Flowchart of the antenna radiation algorithm with the frequency loop. Scripts containing frequency loop are grayed.

is used for the visualization of selected antenna characteristics (surface current distribution, radiation intensity distribution, and directivity patterns) at a single frequency. The desired frequency within the bandwidth should be specified at the beginning of each code. Note that this frequency is not necessarily exactly equal to the grid frequency: the nearest frequency of the grid is used to create the output.

If necessary, the scripts rwg3.m and efield2.m, containing frequency loops, can be converted to executable files, using Matlab compiler, as described in Chapter 2.

7.3. ANTENNA STRUCTURES UNDER STUDY

Four antenna structures will be investigated in this chapter: dipole, bowtie, spiral, and a fractal antenna. We will see that the dipole itself is not a broadband but a narrowband antenna. Two "classic " members of the class of broadband antennas include the bowtie and the spiral antennas. The fractal antenna (Sierpinski gasket) is also not a broadband but rather *a multi-band antenna*.

The antenna meshes are generated by relatively short Matlab scripts bowtie.m, spiralplane.m, and fractal.m in the subdirectory mesh of the root directory Matlab. Before performing any calculations, you may want to run these scripts. The scripts allow for a wide range of antenna parameters. The script bowtie.m can further be modified to create other shapes of planar antennas. The script spiralplane.m is straightforwardly extended to conical spirals.

Three antenna structures are shown in Fig. 7.2. The first one shown in Fig. 7.2*a* is the conventional dipole strip2.mat (244 triangles; 1:100 geometry ratio; 2 m total length) discussed in Chapter 4. The dipole plane is the *xy*-plane. The second structure shown in Fig. 7.2*b* is the bowtie antenna (336 triangles, $\alpha = 90°$ flare angle, 0.2 m total height). The bowtie plane is the *xy*-plane. The structure is created using the script bowtie.m in the subdirectory mesh. The operation of this script utilizes Delaunay triangulation (Matlab function delaunay) as will be discussed in Section 7.7.

The next antenna structure is a plane Archimedean spiral shown in Fig. 7.2*c* (see also [1, pp. 545–550] and [2, pp. 281–287]). The structure has 300 triangles and is 0.2×0.2 m in size. It is created using the script spiralplane.m in the subdirectory mesh. The operation of this script utilizes a deformation of the already pregenerated strip mesh (the script strip.m in subdirectory mesh), and it will be discussed in some detail in Section 7.11. The Archimedean spiral follows the dependence

$$r = r_0 + b\phi \qquad (7.3)$$

in terms of polar coordinates r,ϕ in the *xy*-plane. Archimedean spirals accrue in constant steps and require radiation conductors of constant thickness. They have been regarded as a simpler solution to replace the logarithmic spiral [2, p. 285]. Scripts strip.m and spiralplane.m allows us to vary the spiral

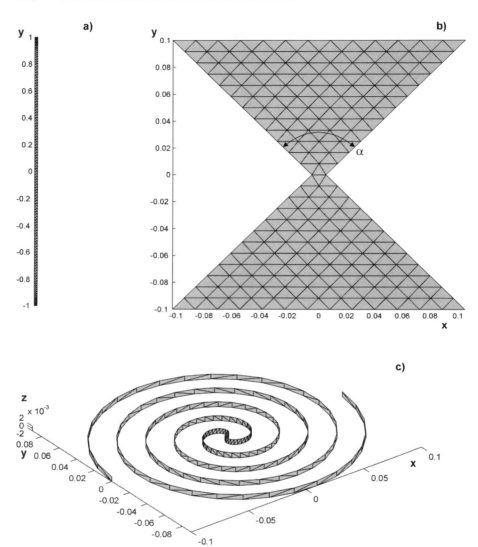

Figure 7.2. Three antenna geometries under study. The feeding edge is always at the origin.

step, number of turns, discretization accuracy, and the width of the corresponding strip.

The Sierpinski fractal bowtie antenna [14,15] is shown in Fig. 7.3 for the first four stages of growth. The parameters used in the script fractal.m are $S = 1, 2, 3, 4$ and $SM = 7$. Here S is stage of fractal growth and SM is stage of mesh growth. We construct this fractal by an operation that excises an inverted

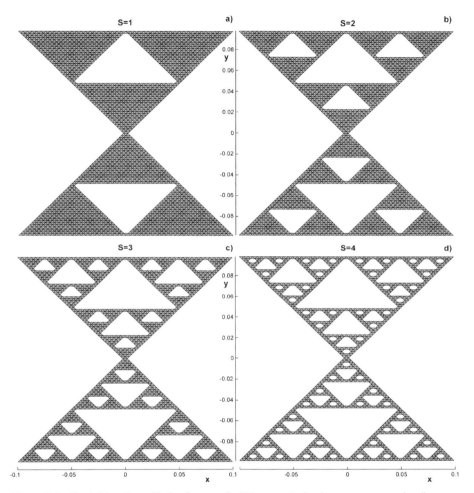

Figure 7.3. Fractal bowties with the flare angle 90° generated using fractal.m. (a–d) correspond to stages of growth which are S = 1, 2, 3, 4.

initial triangle scaled by one half (the generator). The result is shown in Fig. 7.3a. Application of the generator for the first time leaves three smaller filled triangles, to which we may apply the scaled copy of the generator again. The corresponding result is shown in Fig. 7.3b. We iterate this process one more step to obtain Fig. 7.3c, and so on.

Note that the key MoM script rwg3.m always identifies the feeding edge of the antenna as the one closest to the origin. Therefore all antenna structures must have the center of the feeding edge close to the origin. The script rwg3.m should be modified if other antenna geometries are considered.

7.4. DIPOLE IMPEDANCE AND POWER RESONANCE

To obtain frequency characteristics of the dipole, we specify the mesh file name as `strip2.mat` at the top of script `rwg1.m`. Then scripts `rwg1.m`, `rwg2.m` are executed. Frequency loop parameters are specified in script `rwg3.m`. For the present dipole, with the total length of 2 m, we choose

$$f_L = 25\,MHz, \quad f_U = 500\,MHz \tag{7.4}$$

and `NumberOfSteps=200` in order to highlight a few first resonances. Here `NumberOfSteps` is the number of sampling points over the bandwidth. Then the script `rwg3.m` is executed. The final result is observed using the script `sweeplot.m` and is shown in Fig. 7.4. Run this script to see the already precalculated results.

Note that calculations for 200 sampling points may be rather lengthy. For the dipole with 244 triangles, about 9 min CPU running time is necessary to run script `rwg3.m` with 200 frequency steps (on a Pentium IV processor with 1.7 GHz clock speed). Therefore it makes sense to convert the script `rwg3.m` to a function of frequency and use Matlab compiler as was done in Chapter 2.

From Fig. 7.4 we see that the input impedance of the dipole is subject to large oscillations, both real and imaginary parts, when frequency of the feed voltage increases. The input impedance of the center-fed dipole is purely resistive for certain lengths, called *resonant lengths*. The bar in Fig. 7.4a and b denotes the position of the half-wavelength dipole (close to the lowest resonance). The resonant lengths occur at approximately (see Section 4.5 of Chapter 4).

$$2h = \frac{\lambda}{2}, \frac{3\lambda}{2}, \frac{5\lambda}{2}, \dots$$

The dipole may operate at one of the lowest resonances, within the bandwidth of 8 to 16% ([2], p. 260).

To answer the question why the *resonant frequencies* are so popular in electrical communications, we begin by analyzing the radiated (output) power of the dipole antenna. The radiated power has a maximum when the antenna impedance is purely real. Thus the dipole antenna fed by an ideal voltage source becomes the most efficient radiator at the resonant frequencies. The next section provides the quantitative results for radiated power.

The slope of the impedance curves in Fig. 7.4 is changed when one changes the dipole thickness (or the width of an equivalent strip; see Chapter 4). The general rule is a decrease of the peak impedance magnitudes with an increase of thickness. Thus thicker dipoles have a more smooth impedance curve and are in that sense more "broadband." Thick cylindrical dipoles are studied in Ref. [1, ch. 9]. Note that a comprehensive study of different dipole

DIPOLE IMPEDANCE AND POWER RESONANCE 159

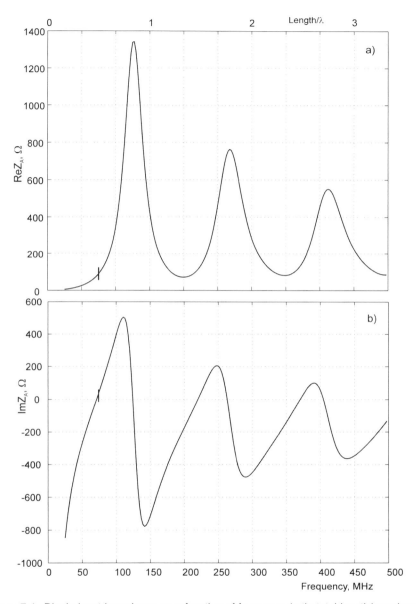

Figure 7.4. Dipole input impedance as a function of frequency (ratio total length/wavelength). A 2 m long and 0.02 m wide strip is considered. The bar indicates the half-wavelength dipole.

7.5. DIPOLE RADIATED POWER, RETURN LOSS, AND GAIN

To obtain the total radiated power and gain as a function of frequency, the script efield2.m should be executed after the script rwg3.m. That script has the same frequency loop, which involves the calculation of radiation intensity distribution and radiated power over a large sphere with 500 triangles (sphere.mat). The script efield2.m is faster than rwg3.m.

The result is observed using the script sweeplot.m and is shown in Fig. 7.5a. We see that the radiated power has strong peaks that correspond to the resonant frequencies discussed above. The dipole reactance crosses zero (becomes negligibly small) exactly at the resonant frequencies. However, the power peaks may be slightly shifted due to finite dipole thickness and a non-sinusoidal current distribution along the dipole. For example, Chapter 4 predicts the total radiated power of 4.5 mW for the half-wavelength dipole. Figure 7.5a shows that there exists a more favorable frequency, slightly lower than the half-wavelength frequency of 75 MHz. At that lower frequency, the dipole will radiate as much as 7 mW.

An ideal voltage generator can be assumed in the feed by a simple power consideration. Usually this voltage generator occurs in series with a resistance (50 Ω). Therefore the more practical "resonant" parameter is the *return loss* *RL* of the antenna. The return loss is found based on the reflection coefficient, Γ, in the antenna feed versus the 50 Ω transmission line, namely

$$\Gamma = \frac{Z_A - 50\,\Omega}{Z_A + 50\,\Omega} \quad (7.5)$$

where Z_A is the antenna input impedance. The discussion of the reflection coefficient will be continued in the next chapter. The return loss is simply the magnitude of the reflection coefficient in dB, i.e.

$$RL = 20\log_{10}|\Gamma| \quad (7.6)$$

The return loss can be taken with negative sign as well. The return loss is the most important parameter with respect to the load matching. It characterizes the antenna's ability to radiate the power instead of reflecting it back to the generator. The antenna's bandwidth is often defined as the band over which the return loss is sufficiently small (below −10 dB).

To find the return loss, we only need the antenna's impedance. The corresponding calculation is done in the script sweeplot.m. The result is shown in Fig. 7.5b. The antenna's resonances are seen as deep notches on

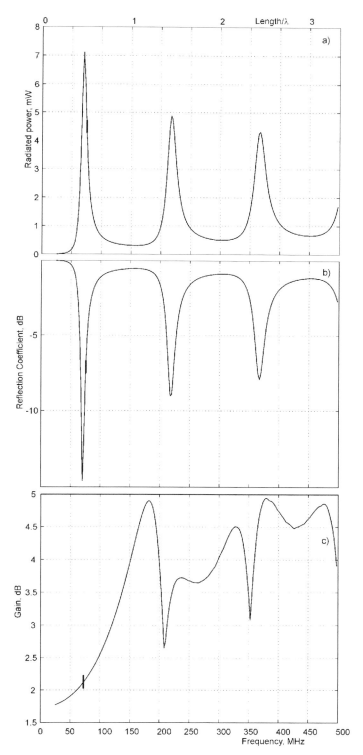

Figure 7.5. Dipole radiated power, return loss, and gain as a function of frequency (ratio total length/wavelength). The bar indicates the half-wavelength dipole.

the plot. Interestingly the resonances in Fig. 7.5a and b are practically identical.

The slope of the power curve and the RL curve in Fig. 7.5a and b become smoother when one increases the dipole thickness (or the width of an equivalent strip; see Chapter 4). The general rule is a decrease of the power peaks with an increase of thickness.

Finally, Fig 7.5c shows the dipole gain as a function of frequency. There is no correlation between the gain and the antenna's input impedance or radiated power. At the same time the slope of the gain curve is in a very good agreement with the corresponding analytical result of Ref. [1, p. 158] even though that result was obtained for an infinitesimally thin dipole. In particular, the gain magnitude of the first maximum differs by 0.25 dB only. At high frequencies the discretization accuracy of the structure `sphere.mat` is not enough to correctly predict the gain values. Instead, `sphere1.mat` should be used.

In conclusion, the conventional dipole should be treated as a typical resonant antenna. It has strongly oscillating input impedance and gain in the frequency band $1/3 \leq 2h/\lambda \leq 3$ and cannot be treated as a broadband antenna.

7.6. DIPOLE COMPARISON WITH NEC MODELING

For the purpose of comparison, we present here simulation results for an equivalent dipole, obtained using a NEC wire solver. SuperNEC software of Poynting Software Pty Ltd. [17] is employed to model a 2 m long wire dipole with an equivalent radius of 0.005 m. The wire is divided into 19 and 39 segments. Option "thin wire kernel" is used.

Figure 7.6 shows the real part of the input impedance (input resistance) compared to the NEC solver. The NEC data are denoted by stars. At low frequencies, smaller than or equal to the half-wavelength frequency, the results come in close proximity to each other. When the frequency increases, the difference becomes more apparent.

This difference is clearly reduced by increasing the number of segments in the wire model as shown in Figs. 7.6b and 7.7b. The problem is, however, that the number of segments in the NEC model is limited by the condition that requires a segment of a length-to-radius ratio greater than about 10 [18]. If we use a wire with 79 segments, instead of 19 or 39 segments, the results will be more accurate at high frequencies. This is shown in the last column of Table 7.1. Simultaneously, the results will become more inaccurate (!) at low frequencies (second column of Table 7.1).

The dilemma of segment count can be solved in NEC by taking two different wire models: one with a small number of segments at low frequencies and another, with a large number of segments at high frequencies. Such an

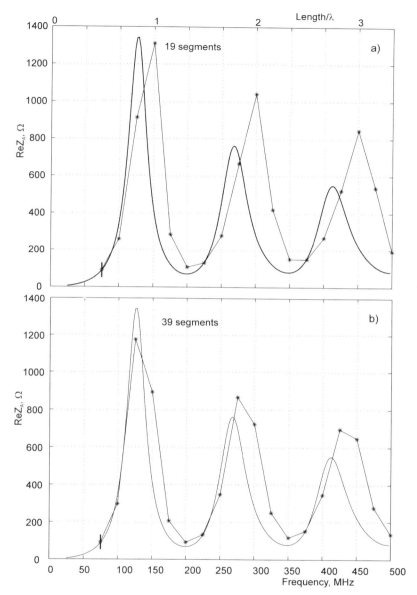

Figure 7.6. Input resistance of the dipole as a function of frequency. Stars denote the NEC results for 19- and 39-segment wires. The bar indicates the half-wavelength dipole.

experiment is rather inconvenient though. Note that the strip model is free of these restrictions. Indeed, at very high frequencies, when the strip width becomes a relatively large fraction of wavelength, the difference between the cylinder model and the strip model would be of a physical nature.

164 BROADBAND ANTENNAS: THE FREQUENCY SWEEP

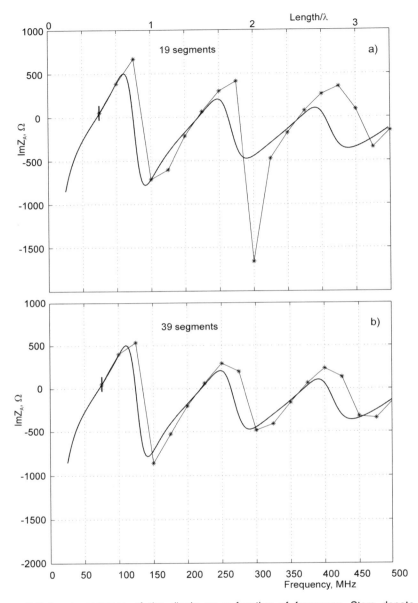

Figure 7.7. Input reactance of the dipole as a function of frequency. Stars denote the NEC results for 19- and 39-segment wire, respectively. Bar indicates the half-wavelength dipole.

MATLAB MESH FOR BOWTIE ANTENNA USING DELAUNAY 165

Table 7.1. Input Impedance of Strip and Wire Models of the 2 m Long Dipole

Model	Input Impedance, Ω at 75 MHz	Input Impedance, Ω at 497.625 MHz
Strip2.mat—244 triangles	**87.6 + j × 47.3**	**82.3 − j × 135.6**
Wire with 19 segments	**89.9 + j × 52.7**	208.7 − j × 180.1
Wire with 39 segments	92.2 + j × 50.2	140.1 − j × 156.0
Wire with 79 segments	97.1 + j × 137.9	**118.1 − j × 148.5**

Note: The strip width is 0.02 m. The equivalent wire radius is 0.005 m.

7.7. MATLAB MESH FOR BOWTIE ANTENNA USING DELAUNAY

To obtain frequency characteristics of the bowtie antenna, we specify the mesh file name as `bowtie` at the top of the script `rwg1.m`. Further the calculations are identical to these for the dipole. Before discussing them, we describe briefly the mesh generation script `bowtie.m` in the subdirectory `mesh`. This script does not use Matlab PDE toolbox. Instead, we use Delaunay triangulation (Matlab function `delaunay`) to create the antenna mesh. The first step is to create 2D node points on the bowtie boundary and in the inner domain. These points are saved in Matlab arrays `X` and `Y`. The conventional array of Cartesian node coordinates, `p(1:3,:)`, is obtained by

```
p=[X; Y; zeros(1,length(X))];
```

Next we call `delaunay` to create the triangle array `t(3,:)`, that is, to identify *nonintersecting* triangles of the mesh as follows

```
TRI = delaunay(X,Y); t=TRI';
```

Finally, we add the fourth row to the triangle array (the domain number 1 is default)

```
t(4,:)=1;
```

The antenna mesh, containing `p` and `t`, is thus defined. A problem with the bowtie structure is its nonconvex shape. Therefore the function `delaunay` creates some nonphysical triangles outside the antenna's area. To eliminate these triangles, we define a polygon, `Xpol`, `Ypol`, whose shape is the antenna's shape. Then array `Center(1:3,:)` of the triangle's center points is created. The line

```
IN = inpolygon(Center(1,:), Center(2,:), Xpol,Ypol);
```

creates an array IN whose element has value 1 if the corresponding triangle (its center) is inside the polygon. Using array IN, the nonphysical triangles are removed from the mesh. Run script bowtie.m to see the corresponding operation sequence displayed as a number of 2D plots. The present algorithm can be applied to create other planar antenna shapes.

7.8. BOWTIE IMPEDANCE

Scripts rwg1.m, rwg2.m are executed first. Since the bowtie length is 10 times smaller than the length of the dipole, the frequency band from 25 MHz to 5 GHz is chosen in the script rwg3.m. After the script rwg3.m is executed the script sweeplot.m outputs the data for the input impedance of the bowtie shown in Fig. 7.8 by a solid line. To obtain the radiated power and gain, we have to run efield2.m in addition to rwg3.m.

The dashed line in Fig. 7.8 shows the impedance values corresponding to a dipole of the same length 20 cm. To obtain these values, mesh strip2.mat was used, with the array p multiplied by 0.1. One can see that the impedance characteristics of the bowtie are considerably smoother than the dipole characteristics. Especially inviting is the behavior of the reactance, which has nearly constant negative value (capacitive reactance) in the band 1.5 to 5 GHz. To further proceed with the impedance analysis (calculation of return loss), we can use the script sweeplot.m.

The impedance behavior of the bowtie is considerably affected by the value of the flare angle, α. Problems at the end of the chapter give a few examples of the bowtie antenna behavior at different flare angles. The classic experimental paper on conical and bowtie antennas is that of Brown and Woodward, RCA Review, 1952. The use of RWG edge elements for the bowtie antenna analysis was discussed yet in Ref. 11 of Chapter 4.

7.9. BOWTIE RADIATED POWER AND GAIN

To obtain the total radiated power and gain as a function of frequency, the script efield2.m should be executed after script rwg3.m. This script saves radiated power (variable TotalPower) and gain (variable GainLogarithmic or GainLinear) in the binary output file gainpower.mat that is used as an input to sweeplot.m. The result is observed using the script sweeplot.m, where these arrays are plotted versus frequency (array f), and is shown in Fig. 7.9.

Only one power resonance can be identified for the present bowtie with $\alpha = 90°$. This resonance occurs when the antenna length/wavelength ratio is slightly smaller than 1:3. The first resonance is very strongly developed and a high output power of about 15 mW is observed at the resonance. The bowtie

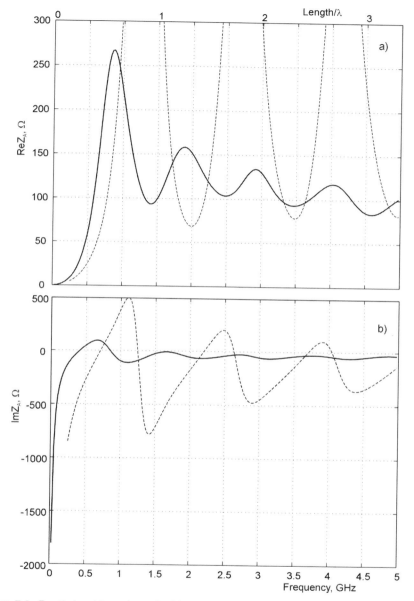

Figure 7.8. Bowtie input impedance (solid line) as a function of frequency (ratio of total length/wavelength). The flare angle is 90°. The dashed curve gives the impedance of the dipole of the same length.

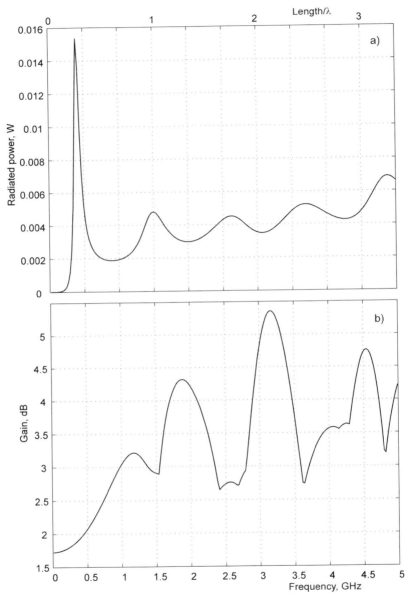

Figure 7.9. Bowtie radiated power and gain as a function of frequency (ratio total length/wavelength). The flare angle is 90°.

gain varies from approximately 2 to 5 dB and is on the order of the dipole gain. When the frequency increases, the output power exhibits moderate oscillations and generally increases as well.

Whereas for the dipole antenna the harmonic current is reflected back and forth from the dipole ends, for the bowtie antenna the current is distributed over a larger surface and eventually dissipates into the radiated field before it reaches the antenna's end. The same situation happens for a thick cylindrical dipole or a sheet dipole. Therefore reflections from the antenna's end are significantly reduced, which makes the bowtie antenna an essentially nonresonant structure, attractive for both broadband and UWB (see Chapter 8) applications.

7.10. BOWTIE RADIATION INTENSITY DISTRIBUTION

After the two frequency loops (rwg3.m and efield2.m) are complete, scripts rwg5single.m, efield2single.m, and efield3single.m display the antenna parameters at any frequency within the bandwidth. The desired frequency should be specified at the beginning of each code. Note that the desired frequency is not necessarily equal to the grid frequency: the nearest frequency of the grid is used to create the output. In this section we are interested in the radiation intensity distribution for the bowtie (script efield2single.m) and the corresponding directivity patterns (script efield3single.m). The output of efield2single.m (3D radiation patterns at several frequencies) is shown in Fig. 7.10. The sphere radius of 1000 m is chosen.

Figure 7.10 indicates that despite the better impedance behavior, the present bowtie antenna might have a diverse radiation intensity distribution through the band 1 to 4 GHz. Note that the color bar in Fig. 7.10 extends from the maximum to minimum intensity magnitude. Therefore relative variations of the intensity may be not as high as they appear in Fig. 7.10. An additional test is necessary to quantitatively describe the intensity distribution.

For this purpose we apply the script efield3single.m, which outputs the directivity patterns in dB. The xz-plane is chosen for comparison. The output of the script is shown in Fig. 7.11 where we plot four directivity patterns corresponding to 1, 2, 3, and 4 GHz. The x-axis corresponds to the reference angle zero.

It is seen in Fig. 7.11 that the results at 1 and 2 GHz indicate a nearly omnidirectional (strictly speaking, *bidirectional*; see [2]) radiation pattern, with relatively small directivity variations. Interestingly the direction of maximum radiation changes from the z-direction in Fig. 7.11a to the x-direction in Fig. 7.11b. The radiation patterns at lower frequencies are also omnidirectional and are quite similar to that shown in Fig. 7.11a. However, the results at 3 and 4 GHz are hardly acceptable. This circumstance, combined with the impedance's behavior, allows us to identify the bandwidth of the present antenna type as approximately 0.5 to 2 GHz.

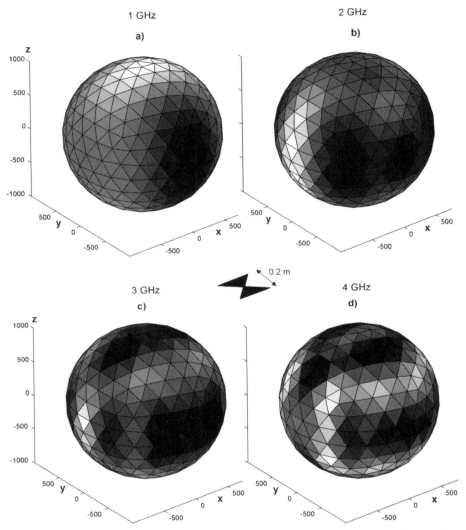

Figure 7.10. Radiation intensity distribution of the bowtie antenna at four different frequencies. The flare angle is 90°.

Further work may be needed to increase the bandwidth. Some of the methods include resistive loading [7,8], dielectric coating [9,10], use of a cavity-backed bowtie [11], and the use of a volume absorber [12]. One other possible method is further optimization of the shape of the metal sheet bowtie antenna. This task may be accomplished numerically.

Note that the bowtie geometry studied here is straightforwardly expanded to one of the simplest fractal antennas: the so-called Sierpinski fractal antenna

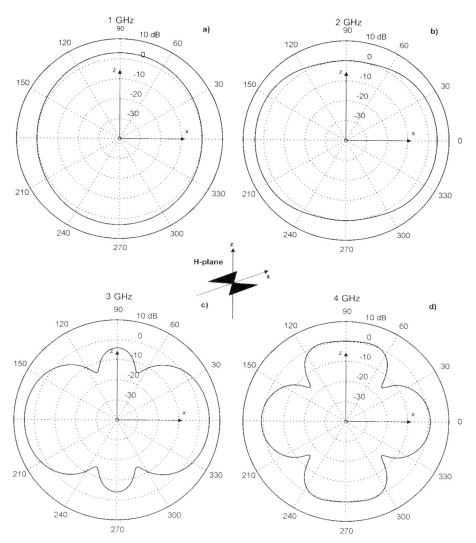

Figure 7.11. Radiation patterns of the bowtie antenna at four different frequencies in the xz-plane.

or Sierpinski gasket [14,15]. Very basically, the Sierpinsli fractal is none other than a bowtie with a specially arranged hole structure. Fractal antennas are not broadband in the conventional sense but rather "multi-band." These antennas are able to keep nearly the same performance at several (up to five) relatively narrow bands [15]. At the end of the chapter we will consider an analysis of the fractal bowtie antenna.

7.11. MESH FOR A SPIRAL ANTENNA

The mesh for a spiral antenna is created using two scripts: strip.m and spiralplane.m in the subdirectory mesh. The first script only creates a thin plane strip. In contrast to Chapter 4, the strip is created in the *xz*-plane. The script spiralplane.m imports the strip geometry from the binary file strip.mat and "bends" it to obtain an Archimedean spiral (see Section 7.3). Since the corresponding code sequence is very short, we present here the complete source code in the form:

```
clear all
load strip
N=5;            %Number of turns
Size=0.2;       %Spiral size in m
angle=abs(2*pi*N*p(1,:));

P(1,:)=p(1,:).*cos(angle);
P(2,:)=p(1,:).*sin(angle);
P(3,:)=p(3,:);
p=Size*P;

save spiralplane p t;
viewer('spiralplane')
```

Variable Size defines the total size of the spiral in its plane. Running scripts strip.m and spiralplane.m in its present form leads to the spiral of five turns shown in Fig. 7.12a (or Fig. 7.2c). If we want to increase the number of turns, the maximum edge size of the mesh should be reduced. To do so, we run the script strip.m using finer discretization, say, Ny=300. This corresponds to 600 triangles in the mesh. Then the script spiralplane.m is executed with a larger number of turns, say, N=10. This leads to the spiral shown in Fig. 7.12b. The script outputs the binary file spiralplane.mat used as an input to the main code sequence. The feeding edge is always located at the origin. The spiral meshes may have as large as 3000–3500 triangles (spirals with 50–75 turns). These meshes can be deformed to form a conical spiral antenna (see [1, pp. 549–550]).

7.12. SPIRAL ANTENNA'S IMPEDANCE, POWER, AND GAIN

To obtain the frequency characteristics of the spiral antenna, we specify the mesh file name as spiralplane at the top of the script rwg1.m. Then scripts rwg1.m, rwg2.m are executed. Since the spiral size is equal to the size of the bowtie antenna from the preceding sections, the frequency band in the script rwg3.m is chosen to be the same, from 25 MHz to 5 GHz. After running the script rwg3.m, script sweeplot.m outputs the data for the input impedance of the spiral antenna shown in Fig. 7.13. These data correspond to the spiral shown in Fig. 7.12a.

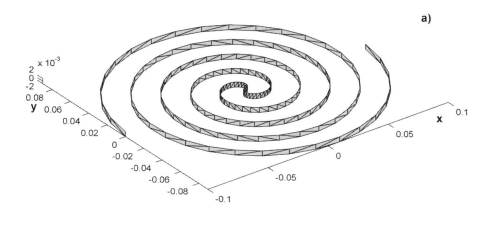

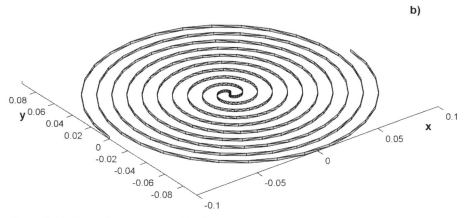

Figure 7.12. Two spiral antennas with (a) 5 turns and (b) 10 turns. The number of triangles is 300 and 600, respectively.

One can see in Fig. 7.13 that the input impedance becomes very flat when frequency exceeds approximately 0.5 GHz. In that area the input resistance varies around 200 Ω. This value is close to the theoretical prediction of 188.5 Ω for equiangular (exponential) spirals (see [2, pp. 281–287] or [1, pp. 545–548]). The input reactance in Fig. 7.13b tends to a negative constant value of approximately −150 Ω when frequency increases above 0.5 GHz. This observation is also in line with the corresponding data for equiangular slot spirals (see [1, p. 549]). Compared to the bowtie antenna with a flare angle of $\alpha = 90°$, the spiral antenna of the same size exhibits smaller variations of the input impedance in the frequency range $f \geq 0.5$ GHz, or, which is the same, when the antenna size is greater than or equal to one-third the wavelength.

174 BROADBAND ANTENNAS: THE FREQUENCY SWEEP

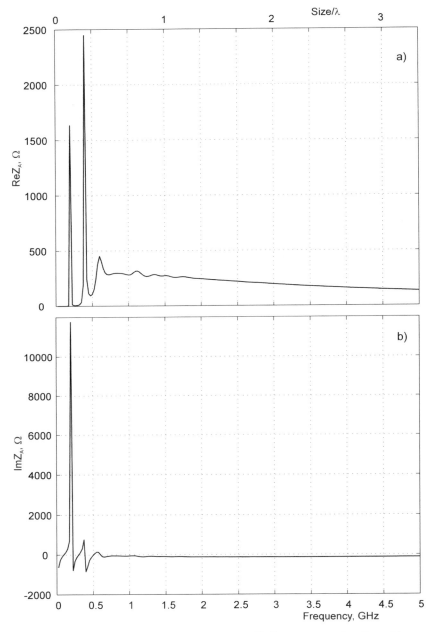

Figure 7.13. Input impedance of the spiral antenna with five turns as a function of frequency (ratio of total length/wavelength). The antenna size is 0.2 m.

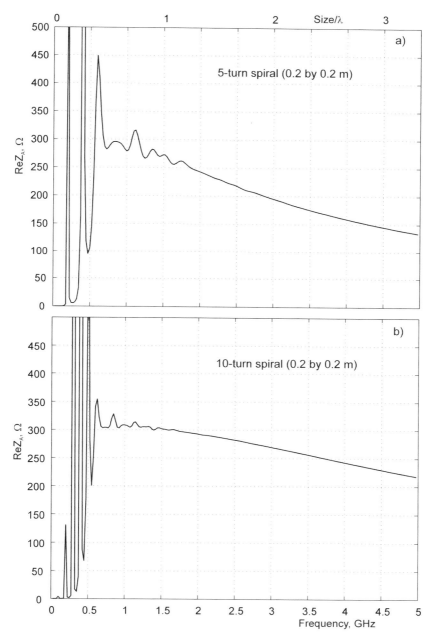

Figure 7.14. Enlarged input resistance (solid line) as a function of frequency (ratio of total length/wavelength). (a) 5-turn spiral antenna; (b) 10-turn spiral antenna. The antenna size is 0.2 m.

A better resolution of impedance behavior at high frequencies is shown in Fig. 7.14a with the data for the input resistance of the 5-turn spiral antenna provided on a larger scale. In order to see how the number of turns affects impedance, corresponding data are presented in Fig. 7.14b for a 10-turn spiral antenna of the same size as was depicted in Fig. 7.12b. The impedance behavior in Fig. 7.14b appears a little smoother except for extremely large oscillations at low frequencies. The magnitude of the input resistance slightly increases with increasing the number of the turns. Similar results are observed for the input reactance.

Figure 7.15 shows simulation results for the total radiated power and the antenna gain in the case of the 5- and 10-turn spiral antennas, respectively. The solid line indicates the solution for the 5-turn spiral antenna, and the dashed line corresponds to the 10-turn spiral. The power behavior of two antennas is quite similar. The 10-turn antenna has smaller power variations in the range 0.5 to 5 GHz.

Spiral antennas are usually cavity backed with an absorber or with a ground plane [19–22]. They are intensively investigated right now due to their inviting characteristics such as broad bandwidth (up to 10:1), circular polarization, and a small physical size. To investigate the polarization properties in the far and near field of the antenna, the script efield1single.m can be used.

In spiral antennas most radiation comes from the region of the structure where the circumference is about one wavelength, often called the *active region*. Thus, as frequency changes, different parts of the spiral antenna will support the majority of the current. This feature is responsible for broadband performance [2, p. 287]. An antenna with distinct active regions is usually very small and essentially appears as if it were infinite. The script rw5single.m allows us to visualize current distribution along the spiral at different frequencies. A study made for two present spirals showed, however, that the active region is usually not very well developed and sometimes cannot be identified properly.

7.13. SPIRAL ANTENNA'S RADIATION INTENSITY DISTRIBUTION

Once two frequency loops (rwg3.m and efield2.m) for the spiral antenna are complete, we can again use scripts efield2single.m and efield3single.m to evaluate antenna parameters at any single frequency within the bandwidth. The output of the script efield2single.m (3D radiation patterns) is shown in Fig. 7.16 at 1, 2, 3, and 4 GHz. The sphere radius is 1000 m.

Compared with the 3D intensity patterns for the bowtie antenna of Fig. 7.10, the spiral antenna has a considerably smoother radiation pattern evolution when frequency increases. The pattern remains bidirectional in the entire frequency band from 0.5 GHz to 4 GHz. To support this conclusion, we present in Fig. 7.17 three directivity patterns of the spiral antenna in the xz-plane. The x-axis corresponds to the reference angle zero. These patterns are obtained

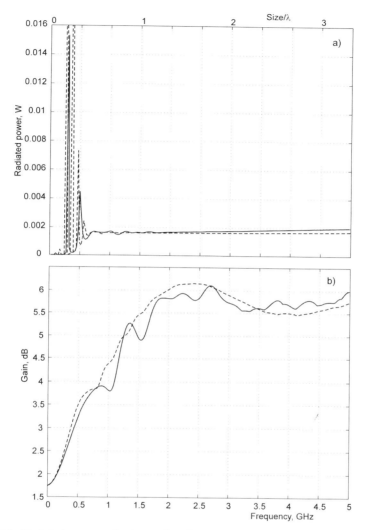

Figure 7.15. Radiated power and gain as a function of frequency. Solid line: Spiral with 5 turns; dashed line: spiral with 10 turns. The antenna size is 0.2 m.

using the script efield3single.m. It is seen in Fig. 7.17 that the bidirectional pattern remains very stable when frequency increases.

A cavity-backed spiral antenna or a spiral over a ground plane would create unidirectional radiation (single lobe). A gain up to 14 dB is obtained for a spiral covered by a dielectric layer [22]. This value by far exceeds the gain found in Fig. 7.15. The unidirectional radiation pattern is also observed for a conical spiral antenna [1, pp. 549–550].

178 BROADBAND ANTENNAS: THE FREQUENCY SWEEP

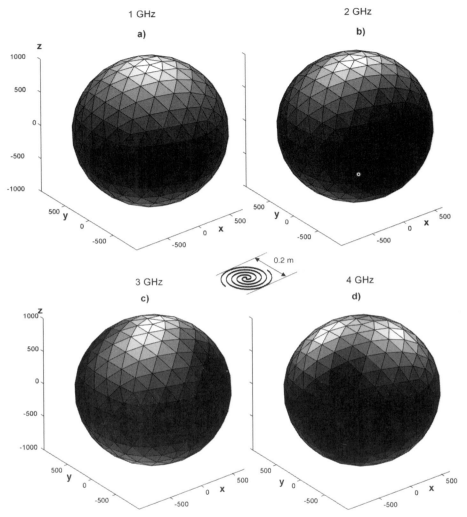

Figure 7.16. Radiation intensity distribution of the spiral antenna with five turns at four different frequencies. The antenna size is 0.2 m.

The spiral antennas illustrate the principle that emphasis on angle-related geometry rather than on the length-related one (e.g., the dipole) will lead to broadband antennas. Another and very important type of broadband directional antennas are log-periodic antennas and arrays (see [1, sec. 11.4] and [2, ch. 6]). Log-periodic antennas are ideologically similar to spiral antennas. However, it is more difficult to create their shapes using the corresponding Matlab scripts. An alternative is to employ the Matlab PDE toolbox discussed in previous chapters.

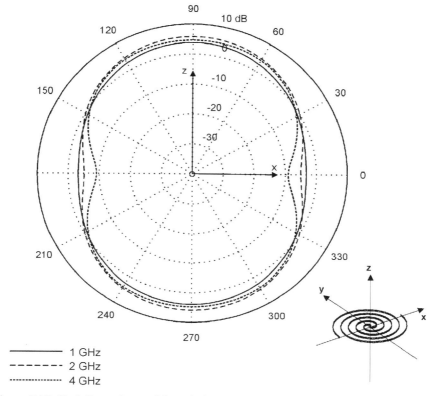

Figure 7.17. Radiation patterns of the spiral antenna with five turns at three different frequencies in the xz-plane. The antenna size is 0.2 m.

Script `bowtie.m` could, in principle, be extended to create toothed log-periodic antennas [1, figs. 11.6, 11.7]. As concerns the log-periodic array, Chapter 6 provides all necessary tools to create a log-periodic array of dipoles (the end-fire array with the phase shift of 180° and with logarithmically increasing spacing).

7.14. MULTIBAND ANTENNAS: THE SIERPINSKI FRACTAL

In order to create a bowtie shaped as a Sierpinski gasket (Fig. 7.3), we use the Matlab script `fractal.m` from the subdirectory `mesh`. That script was tested with the stages of fractal growth $S = 1, 2, 3, 4$. First, half of the bowtie structure (the initiator triangle) is created. The algorithm of the script is based on applying the generator to a number of filled triangles, at each refinement step. It utilizes the function `divider` that for an arbitrary triangle, outputs the central subtriangle to be cut and three remaining subtriangles. As a result arrays CatPolygonX(1:3,:) and CatPolygonY (1:3,:) are created that identify multiple polygons (triangles) to be cut from the bowtie at each stage. The cutting procedure is similar to that described in Section 7.7. After the nonphysical triangles are removed from the mesh, the mesh is cloned in order to create the symmetrical bowtie. An intermediate operation includes removal of doubled vertexes from the mesh when cloning intercepting structures.

A principal point for the RWG analysis is that the mesh patches at the vertexes of the subtriangles should be kept in the mesh as shown in Figs. 7.3 and 7.18. Otherwise, the current would not be able to flow through the junctions at the boundary and the completely incorrect results would be obtained. This situation is addressed using a special loop in the script fractal.m.

The algorithm of the script fractal.m is very sensitive to the mesh size. Therefore only certain mesh sizes are allowed. They are referred to at the beginning of the script. We will investigate here one fractal antenna, with the flare angle of 90 degrees and $S = 2$, shown in Fig. 7.18. This structure has 857 RWG edge elements. The finer structures shown in Fig. 7.3 require large CPU processor run times and will not be considered in the present chapter. The interested reader could try these structures if the number of sampling points in the frequency domain is relatively small (20–50).

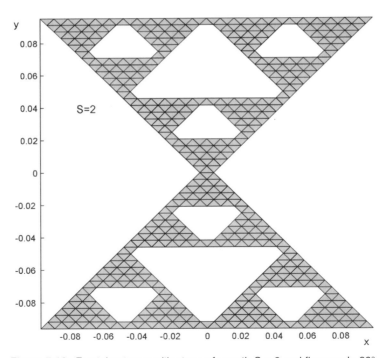

Figure 7.18. Fractal antenna with stage of growth $S = 2$ and flare angle 90°.

7.15. SIERPINSKI FRACTAL'S IMPEDANCE, POWER, AND GAIN

To obtain the frequency characteristics of the fractal antenna, we specify the mesh file name as fractal at the top of the script rwg1.m. Then scripts rwg1.m, rwg2.m are executed. The antenna size is approximately equal to the size of the bowtie antenna in Sections 7.8 to 7.10. The width is the same (0.2 m), whereas the height is somewhat smaller (0.18 m). The difference in the height is due to the different treatment of the bowtie neck. It can easily be corrected if necessary. The frequency band in the script rwg3.m is chosen to cover the range from 0.025 to 8 GHz, totaling 100 sampling points. After running rwg3.m, the script sweeplot.m outputs the data for the input impedance of the fractal antenna shown in Fig. 7.19.

In Fig. 7.19 we compare the input impedance of the fractal antenna (solid line) with input impedance of the bowtie antenna (dashed line). The impedance behavior of the fractal antenna indicates a typical resonant structure. The input resistance has large peaks, whereas the input reactance has multiple nulls.

In our analysis of impedances, the multi-band behavior of the fractal antenna is not yet specified. We are interested in the resonances that are characterized by the zero input reactance and the maximum power radiated by the antenna. Furthermore, if the input resistance at these resonances will be well matched to the 50 Ω load, the multi-band behavior will be established.

The radiated power of the fractal antenna (solid line) and of the equivalent bowtie (dashed line) are shown in Fig. 7.20a. At low frequencies (below 0.5 GHz) both power dependencies are nearly the same. We can see that the fractal antenna has three resonances. The number of resonances is thus the number of fractal stages ($S = 2$) plus one. The first resonance is that of the pure bowtie[1] and corresponds to $f_1 = 0.39$ GHz. This resonance occurs whether or not the fractal structure is present. The second resonance at $f_2 = 1.48$ GHz corresponds to the first fractal iteration, and the third resonance at $f_3 = 3.35$ GHz corresponds to the second fractal iteration. The frequency ratios

$$\frac{f_2}{f_1} = 3.79, \quad \frac{f_3}{f_2} = 2.26, \ldots \qquad (7.7)$$

asymptotically tend to 2 when the number of fractal stages increases as predicted by theory [15]. The resonant bands are thus log-periodically spaced, by a factor of 2, which is exactly the scale factor that related triangle size at each stage of growth.

To investigate the resonances more properly, we check the input reflection coefficient, Γ, of the transmitting antenna given by Eq. (7.5) (script

---

[1] In [15] such a resonance is associated with the first fractal iteration. Calculations done for the antenna from Ref. [15] indicate that this result is rather the resonance of the original monopole bowtie antenna.

182 BROADBAND ANTENNAS: THE FREQUENCY SWEEP

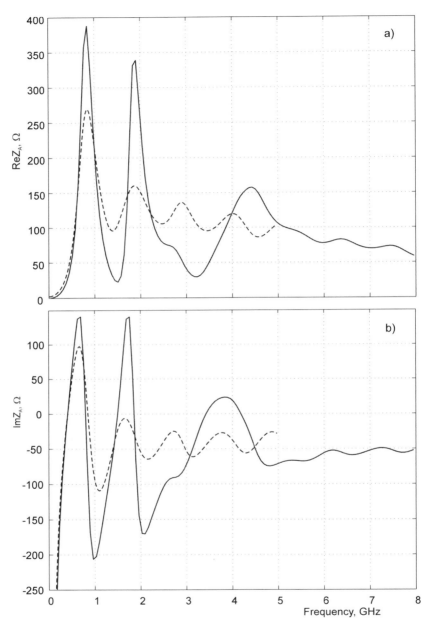

Figure 7.19. Fractal antenna input impedance (solid line) as a function of frequency (ratio of total length/wavelength). Impedance of the equivalent bowtie (dashed line).

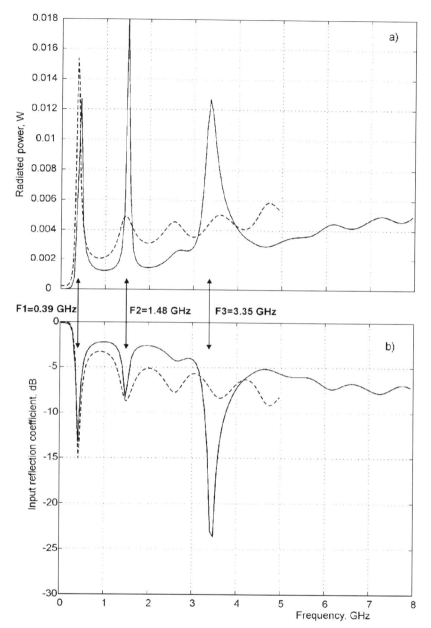

Figure 7.20. (a) Radiated power; (b) input reflection coefficient of the fractal and bowtie antennas. Solid curve: Fractal antenna; dashed curve: equivalent bowtie.

sweeplot.m). Figure 7.20b shows the magnitude of the reflection coefficient to logarithmic scale versus frequency. We see that the power resonances are directly associated with the minima of the reflection coefficient, which enables us to design a well-matched antenna at these frequencies.

The corresponding resonance bands are the multi-bands of the fractal antenna. The relative bandwidth at each band typically reaches 7 to 20% [15]. We note that calculations and experiments of Ref. [15] predict different reflection coefficients for two first resonances in a similar situation. They are on the order of −10 dB (first resonance) and −14 dB (second resonance).

The reason for the different return loss may be a slightly different antenna configuration (the monopole fractal bowtie in [15] has the flare angle of 60°) and the insufficient accuracy of the mesh shown in Fig. 7.18. Nevertheless, a general tendency (reflection coefficient decreases with frequency up to the third resonance) remains exactly the same. Furthermore we obtain a very similar value of the third reflection coefficient.

Finally in this section Fig. 7.21a shows the surface current distribution on the fractal antenna surface at 1.48 GHz (second resonance). The self-similarity of the current distribution [15] can be observed. Figure 7.21b gives two 3D radiation patterns at the second and third resonant frequencies, respectively. Compared to the bowtie radiation patterns in Fig. 7.10, the fractal antenna clearly produces a directional beam of a higher gain at the resonant frequencies.

Notice that the dipole investigated at the beginning of this chapter is also a resonant system. A reasonable question is therefore why not use the simple dipole instead of the fractal antenna for the multi-band purposes. The investigations done in [15] for an equivalent fractal system (the so-called Koch antenna) showed that the fractal antenna improves the bandwidth of each band, the radiation resistance, and the reactance relative to those of a linear monopole (dipole). Equivalently the size of the dipole antenna can be reduced by fractally shaping its linear profile [15].

7.16. CONCLUSIONS

In this chapter we gave four examples of antenna parameter evaluation in the frequency domain, for the dipole, bowtie, spiral, and fractal antenna. The major quantities of interest are the input impedance, radiated power, and gain. The code sequence remains exactly the same for any antenna type; only the input structure's file name has to be changed. Radiation patterns and the surface current distribution at any single frequency within the bandwidth can be examined using the corresponding Matlab scripts with the extension "single."

None of the models of the present chapter require the use of the Matlab PDE toolbox. Instead, we employ either 2D Delaunay triangulation (Matlab

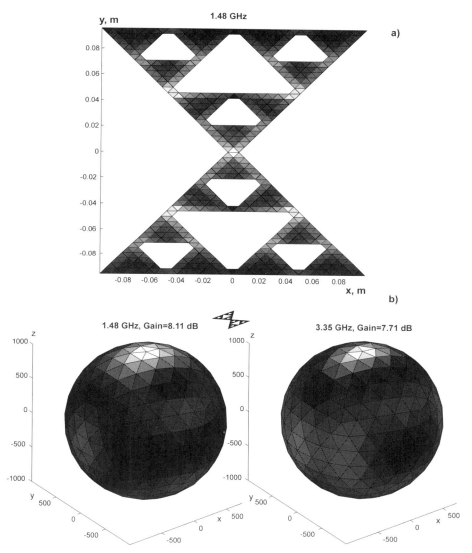

Figure 7.21. (a) Surface current distribution at the second resonance; (b) 3D radiation patterns at the second and third resonances of the fractal antenna.

function `delaunay`) to create the antenna mesh (bowtie antenna) or a custom Matlab script (spiral antenna or fractal antenna). This approach will be explored in the following chapters. By a straightforward modification of the script `rwg3.m`, the algorithm of the present chapter can be extended to antenna arrays.

REFERENCES

1. C. A. Balanis. *Antenna Theory: Analysis and Design*, 2nd ed. Wiley, New York, 1997.
2. W. L. Stutzman and G. A. Thiele. *Antenna Theory and Design*. Wiley, New York, 1981.
3. H. Y. Pao and A. J. Poggio. Design of a TEM waveguide for ultra-wideband applications. In *Antennas and Propagation Society, 1999 IEEE International Symposium*, Vol. 3, IEEE Press, Piscataway, NJ, 1999, 1574–1577.
4. T. T. Wu and R. W. P. King. The cylindrical antenna with nonreflective resistive loading. *IEEE Trans. Antennas and Propagation*, 13: pp. 369–373, 1965.
5. E. Hallén. *Electromagnetic Theory*. Wiley, New York, 1962, at paragraph 35.9 "Reflection-free antennas," pp. 501–504.
6. T. P. Montoya and G. S. Smith. A study of pulse radiation from several broad-band loaded monopoles. *IEEE Trans. Antennas and Propagation*, 44 (8): 1172–1182, 1996.
7. J. G. Maloney and G. S. Smith. Optimization of a conical antenna for pulse radiation: An efficient design using resistive loading. *IEEE Trans. Antennas and Propagation*, 41 (7): 940–947, 1993.
8. K. L. Shlager, G. S. Smith, and J. G. Maloney. Optimization of bow-tie antennas for pulse radiation. *IEEE Trans. Antennas and Propagation*, 42 (7): 975–982, 1994.
9. S. C. Hagness, A. Taflove, and J. E. Bridges. Wideband ultralow reverberation antenna for biological testing. *Electronics Letters*, 33 (19): 1594–1595, 1997.
10. S. C. Hagness, A. Taflove, and J. E. Bridges. Three-dimensional FDTD analysis of pulsed microwave confocal system for breast cancer detection: Design of an antenna array element. *IEEE Trans. Antennas and Propagation*, 47 (5): 783–791, 1999.
11. Y. Nishioka, O. Maeshima, T. Uno, and S. Adachi. FDTD analysis of resistor-loaded bow-tie antennas covered with ferrite coated conducting cavity for subsurface radar. *IEEE Trans. Antennas and Propagation*, 47 (6): 970–977, 1999.
12. E. S. Eide. Ultra-wideband transmit/receive antenna pair for ground penetrating radar. *IEE Proc. Microwave Antennas Propagation*, 147 (3): 231–235, 2000.
13. Y. Chevalier, Y. Imbs, B. Beillard, J. Andrieu, M. Jouvet, B. Jecko, and E. Le Legros. A new broadband resistive wire antenna for ultrawideband applications. In E. Heyman, B. Mandelbaum, J. Shiloh, eds., *Ultra-wideband, Short-Pulse Electromagnetics 4*. Kluwer Academic/Plenum Publishers, New York, 1999, pp. 157–164.
14. D. L. Jaggard, A. D. Jaggard, and P. V. Frangos. Fractal electrodynamics: surfaces and superlattices. In D. H. Werner and R. Mittra, eds., *Frontiers in Electromagnetics*. IEEE Press, New York, 2000, pp. 1–47.
15. C. Puente, J. Romeu, and A. Carmada. Fractal-shaped antennas. In D. H. Werner and R. Mittra, eds., *Frontiers in Electromagnetics*. IEEE Press, New York, 2000, pp. 48–93.
16. R. W. P. King. *Tables of Antenna Characteristics*. IFI Plenum Data Corporation, New York, 1971.
17. *MoM Technical Reference Manual*. Poynting Software Ltd., 2001, 66 pp.
18. D. H. Werner. A method of moments approach for the efficient and accurate modeling of moderately thick cylindrical wire antennas. *IEEE Trans. Antennas and Propagation*, 46 (3): 373–382, 1998.

19. D. S. Filipović and J. L. Volakis. Design and demontration of a novel conformal slot spiral antenna for VHF to L-band operation. In D. H. Werner and R. Mittra, eds., *2001 IEEE Antenna and Propagation Society International Symp.*, Vol. 4. IEEE Press, Piscataway, NJ, 2001, pp. 120–123.
20. M. N. Asfar, Y. Wang, and H. Ding. A new wideband cavity-backed spiral antenna. D. H. Werner and R. Mittra, eds., *2001 IEEE Antenna and Propagation Society International Symp.*, Vol. 4. IEEE Press, Piscataway, NJ, 2001, pp. 124–127.
21. K. Hirose, M. Miyamoto, and H. Nakano. A two-wire spiral antenna with unbalanced feed. D. H. Werner and R. Mittra, eds., *2001 IEEE Antenna and Propagation Society International Symp.*, Vol. 4. IEEE Press, Piscataway, NJ, 2001, pp. 128–131.
22. H. Nakano, Y. Okabe, H. Mimaki, and J. Yamauchi. The effect of an upper dielectric layer on the radiation characteristics of a spiral antenna. D. H. Werner and R. Mittra, eds., *2001 IEEE Antenna and Propagation Society International Symp.*, Vol. 4. IEEE Press, Piscataway, NJ, 2001, pp. 132–135.

PROBLEMS

7.1. For a 2 m long dipole strip2.mat determine the input impedance behavior as a function of frequency in the range 5 to 150 MHz. The sampling interval is 5 MHz. Additionally plot:
   a. Surface current distribution at 150 MHz
   b. 3D radiation pattern at 150 MHz
   c. Directivity pattern in the elevation plane at 150 MHz.

7.2. For a 2 m long dipole strip2.mat determine the input impedance behavior as a function of frequency in the range 0.5 to 1 GHz. The sampling interval is 10 MHz. Additionally plot:
   a. Surface current distribution at 0.5 GHz
   b. 3D radiation pattern at 0.5 GHz
   c. Directivity pattern in the elevation plane at 0.5 GHz.

7.3.* The specified standard for the bandwidth is often variation of the ratio $(1 + |\Gamma|)/(1 - |\Gamma|)$ associated with VSWR (voltage standing wave ratio) at the antenna input terminals. In this case let the specified standard be

$$\text{VSWR} < 2$$

Design a dipole antenna that:
   a. Has the center frequency of 1 GHz
   b. Has the bandwidth of 5%.

7.4.* Repeat task of Problem 7.3 if the center frequency changes to 2 GHz and the bandwidth changes to 10%.

7.5. Create a structure for the bowtie antenna of total length 20 cm, with a flare angle of 60° and a feeding edge length of 1 cm. Determine the input impedance behavior as a function of frequency in the range 25 MHz to 4 GHz. The sampling interval is 25 MHz. Plot:
   a. Surface current distribution at 1 GHz
   b. 3D radiation pattern at 1 GHz.

7.6. Create a structure for a bowtie antenna of total length 20 cm, with a flare angle of 120 ° and a feeding edge length of 1 cm. Determine the input impedance behavior as a function of frequency in the range 25 MHz to 4 GHz. The sampling interval is 25 MHz. Plot:
   a. Surface current distribution at 1 GHz
   b. 3D radiation pattern at 1 GHz.

7.7. Create a structure for the bowtie antenna described in Problem 7.5. Save the script `bowtie.m`, as `bowtie1.m`, and modify that script in order to create a V-shaped bowtie antenna shown in Fig. 7.22. Use `viewer bowtie` to visualize the structure. Determine the input impedance behavior as a function of frequency in the range 25 MHz to 4 GHz. The sampling interval is 25 MHz. Plot the radiation intensity distributions at 1, 2, 3, and 4 GHz. Compare your results with those of Fig. 7.10.

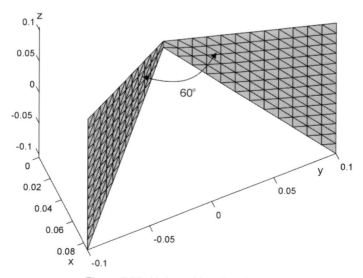

Figure 7.22. V-shaped bowtie antenna.

7.8.* The specified standard for a bandwidth is

$$\text{VSWR} = \frac{1+|\Gamma|}{1-|\Gamma|} < 2$$

where Γ is the antenna's reflection coefficient. Design a bowtie antenna that:
a. Has a center frequency of 3 GHz
b. Has the bandwidth of 3:1.

7.9.* Repeat the task of Problem 7.8 if the center frequency changes to 5 GHz and the bandwidth changes to 6:1.

7.10. Create a structure for the spiral antenna with $N = 2$ turns of total size (diameter) 20 cm. The strip width is 4 mm. Determine the input impedance behavior as a function of frequency in the range 25 MHz to 5 GHz. The sampling interval is 25 MHz. Plot:
a. Surface current distribution at 2 GHz
b. 3D radiation pattern at 2 GHz
c. Directivity pattern in the xz-plane at 2 GHz.

7.11. Create a structure for the spiral antenna with $N = 5$ turns of total size (diameter) 2 cm. The strip width is 2 mm. Determine the input impedance behavior as a function of frequency in the range 0.25 to 50 GHz. The sampling interval is 0.25 GHz. Plot:
a. Surface current distribution at 20 GHz
b. 3D radiation pattern at 20 GHz
c. Directivity pattern in the xz-plane at 20 GHz.

7.12.* Create a structure for the spiral antenna shown in Fig. 7.12a. Save script `spiralplane.m` as `spiralplane1.m` and modify that script in order to obtain the conical spiral antenna shown in Fig. 7.23. Determine the input impedance behavior as a function of frequency in the range 25 MHz to 5 GHz. The sampling interval is 25 MHz. Plot the radiation intensity distributions at 1, 2, 3, and 4 GHz. Compare your results with those of Fig. 7.16.

Hint: Use the line

`p(3,:)=p(3,:)+0.1*(p(1,:).^2+p(2,:).^2)/(Size/2)^2;`

for the structure transformation.

7.13.* Repeat Problem 7.12 for a 10-turn spiral antenna.

7.14. For the fractal bowtie shown in Fig. 7.18, determine the input impedance, surface current distribution, and radiation intensity distribution at three first resonant frequencies.

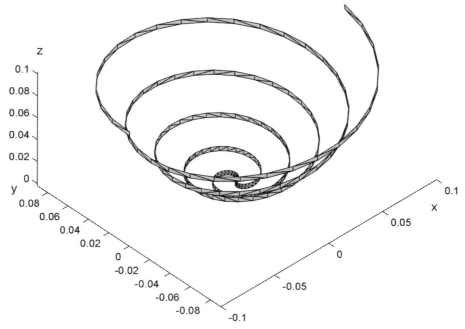
Figure 7.23. Conical spiral antenna with the center feed.

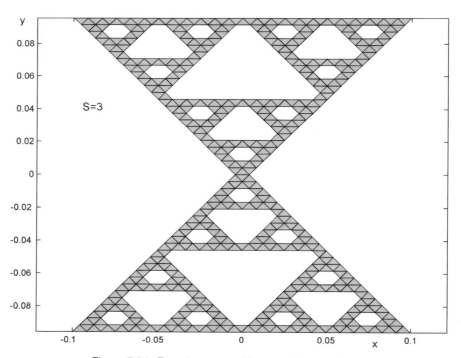
Figure 7.24. Fractal antenna with stage of growth S = 3.

7.15.* Create a fractal bowtie shown in Fig. 7.24 (the same as in Fig. 7.18 but the stage of fractal growth is 3). Assuming a frequency band from 0.025 to 8 GHz totaling 100 sampling points, determine the behavior of the radiated power and the input reflection coefficient in frequency domain. Compare your results with those obtained in Section 7.15 for the previous fractal iteration.

7.16.* Consider a dipole of 20 cm total length. Assuming a frequency band from 0.025 to 8 GHz totaling 100 sampling points, determine the behavior of the radiated power and the input reflection coefficient in frequency domain. Obtain 3D radiation patterns at three resonant frequencies. Compare your results with those obtained in Section 7.15 for the fractal antenna.

8

ULTRA-WIDEBAND COMMUNICATION ANTENNA: TIME DOMAIN ANALYSIS

8.1. Introduction
8.2. Code Sequence
8.3. Incident Voltage Pulse
8.4. Surface Discretization and Feed Model
8.5. Frequency Loop
8.6. Surface Current Distribution
8.7. Antenna Input Impedance
8.8. Antenna Radiation Intensity, Gain
8.9. Directivity Patterns
8.10. Antenna-to-Free-Space Transfer Function
8.11. Antenna-to-Antenna Transfer Function
8.12. Discrete Fourier Transform
8.13. Received Voltage Pulse
8.14. Impedance Mismatch
8.15. Voltage Pulse at a Load
8.16. Conclusions
References
Problems

8.1. INTRODUCTION

This chapter gives a complete simulation of an ultra-wideband pulse slot antenna designed by Time Domain, Co. [1]. The antenna hardware setup is shown in Fig. 8.1. Before reading this chapter you may want to run script rwg34.m to see one of the final results for UWB pulse transmission.

Impulse radiating antennas are members of a class of antennas that are designed for the radiation of ultra-wideband (UWB) electromagnetic pulses, and not for signals at fixed frequencies. For example, small-scale ground penetrating radars (GPR) are designed to operate over a wide frequency range, typically from a few MHz up to 3GHz and higher [2,3]. Ultra-wideband sources and antennas are of interest for a variety of potential applications that range from transient radar systems [2–4] to communication systems (impulse radio [5,6]) and medical diagnostics [7,8]. The antenna shown in Fig. 8.1 is particularly intended for wireless communication links [1,5,6].

For pulse antennas two approaches are currently used to simulate the radiation parameters. One is the time domain method, where all the simulations

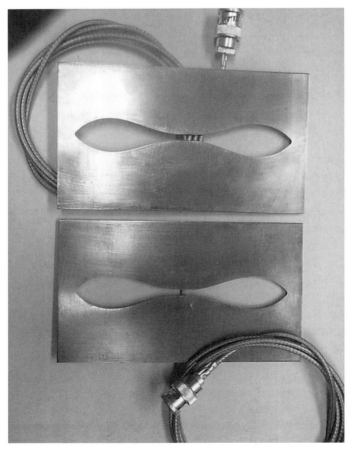

Figure 8.1. Transmitting and receiving slot antennas (copper). Feed cables (partially not seen) are attached to the middle of the slot, as close as possible to the slot's boundary.

are done with time-dependent electric and magnetic fields. The finite-difference time-domain (FDTD) numerical algorithm is the most suitable tool for this purpose [2,8–12]. Another approach is the *frequency domain* method. In this method we calculate antenna performance for every single harmonic wave, within the anticipated pulse bandwidth. The result for a pulse is then obtained using the inverse Fourier transform. The MoM method is ideally suited for the frequency domain analysis [13].

To pursue calculations in the frequency domain; a frequency sweep should be made similar to that discussed in Chapter 7; that is, the antenna parameters should be calculated at a number of frequency samples. We may expect this number to be high if we want good accuracy. At each sample frequency we should know the MoM impedance matrix, surface current distribution, and the input impedance of the antenna. Therefore the moment method becomes both time and space consuming. If only one pulse form is studied, the time domain method will be more suitable and superior to the frequency domain analysis.

At the same time, if we have many different pulses, the computational expenses could nearly be the same. To jump from one pulse form to another the frequency domain method requires simple spectrum multiplication. The time domain method, however, needs the entire simulation process to be repeated. The more different pulse shapes we have, the more advantageous the frequency domain method becomes.

Another important advantage is that the frequency domain method is a straightforward application of the MoM frequency sweep code developed in Chapter 7. The frequency loop of Chapter 7 is extended to calculate the pulse transmission, that is, to derive the antenna-to-antenna transfer function. The transfer function allows us to predict the received voltage pulse, which is the most important parameter of UWB transmission.

8.2. CODE SEQUENCE

The code sequence of this chapter includes three scripts with frequency loops shown in Fig. 8.2. The script rwg31.m is identical to script rwg3.m of the previous chapter. The bandwidth, sampling rate, and the number of sampling frequencies are specified in that script and can be chosen arbitrarily. The script outputs binary file current.mat that contains the antenna impedance and surface current distribution data for every frequency. The second file containing the same frequency loop is the script rwg32.m. This script outputs the radiated electric field (complex vector) at a point as a function of frequency into the binary file radiatedfield.mat. Using this information, we derive an antenna-to-free-space transfer function in frequency domain. The notation rwg32.m is questionable since that script is in fact an extension of the script efield1.m to the frequency loop. Nevertheless, we keep this notation to emphasize the intermediate character of the corresponding transfer function.

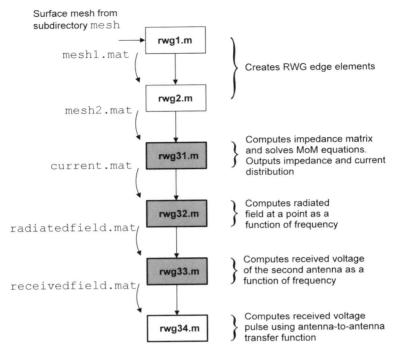

Figure 8.2. Flowchart for the time domain analysis. Scripts containing the frequency loop are grayed.

Next, the script `rwg33.m` solves the scattering problem for a second antenna at every frequency, assuming that the incident field is that due to the first antenna. The antennas may or may not be identical. In that script we calculate the resulting surface current distribution and, using the already known antenna input impedance, determine the received voltage in the feed. The ratio of the received voltage to the input voltage (1 V at the feed point of the transmitting antenna) constitutes the antenna-to-antenna transfer function over the desired frequency band. The transfer function can be modified by introducing an appropriate antenna termination circuit (see below). Once the antenna-to-antenna transfer function is established, the received voltage pulse can be calculated for an arbitrary voltage pulse in the feed of the transmitting antenna, using spectrum multiplication and inverse Fourier transform (script `rwg34.m`).

The present algorithm is not optimal in the sense that the impedance matrix is calculated twice, in the scripts `rwg31.m` and `rwg33.m`. This operation is necessary if we have two distinct antennas but becomes meaningless if a pair of identical antennas is considered. We present here the most general version of the algorithm assuming that the receiving UWB antenna may be different from the base station antenna. For the pair of identical antennas, the scripts `rwg31.m`, `rwg32.m`, and `rwg33.m` can be combined into a single script, with only one impedance matrix at a given frequency. The rest of the codes of this chapter (with extension `single`) are those of Chapter 7.

8.3. INCIDENT VOLTAGE PULSE

A number of "test" pulse forms exist in the RF communication literature. The most popular is the Gaussian monopulse and its modifications [6]. A fundamental characteristic of a monopulse is that it must have zero dc content to allow it to radiate effectively. In other words, the pulse must have equal positive and negative phases so that the total integral over the pulse duration becomes zero. The time domain representation of the typical differentiated Gaussian (originally Rayleigh) monopulse, $p(t, \sigma)$, is [6]

$$p(t, \sigma) = \frac{t}{\sigma^2} e^{-t^2/2\sigma^2} \qquad (8.1)$$

The corresponding frequency spectrum, $P(t, \sigma)$ can be written in the form (see [6])

$$P(f, \sigma) = j\sqrt{2\pi}(2\pi\sigma f)e^{-(2\pi\sigma f)^2/2} \qquad (8.2)$$

The temporal pulse waveform and its spectrum are shown in Fig. 8.3a and b, respectively. In Fig. 8.3 we present results in terms of normalized time, t/σ, and normalized frequency, $f\sigma$. The parameter σ is known as a characteristic time of the pulse.

From Fig. 8.3 we can obtain the following parameters of the pulse:

1. Effective pulse duration (pulse length) [6] (this interval contains 99.99% of the total pulse energy)

$$T = 7\sigma$$

2. Center frequency,[1]

$$f_c = \frac{0.17}{\sigma} = \frac{1.19}{T} \qquad (8.3a)$$

---

[1] The center frequency corresponds to the peak of the pulse spectrum (Fig. 8.3b).

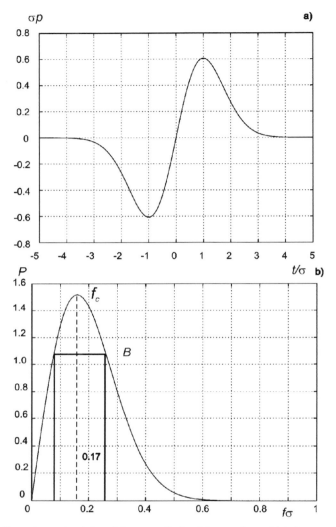

Figure 8.3. Gaussian pulse (a) and the magnitude of its spectrum (b) in terms of dimensionless variables: normalized time, t/σ, and normalized frequency, $f\sigma$.

3. Half-power lower frequency,

$$f_L = \frac{0.09}{\sigma} = \frac{0.63}{T} \qquad (8.3b)$$

4. Half-power higher frequency,

$$f_H = \frac{0.27}{\sigma} = \frac{1.89}{T} \qquad (8.3c)$$

SURFACE DISCRETIZATION AND FEED MODEL   199

5. Half-power (3 dB) bandwidth [6],

$$B = \frac{0.18}{\sigma} = \frac{1.26}{T} \quad (8.3d)$$

For example, take a typical ultra-wideband monopulse of 1 ns duration [1], that is, $T = 7\sigma = 10^{-9}s$. This gives, according to Eqs. (8.3),

$$f_c = \frac{1.19}{T} = 1.19\,\text{GHz}$$

$$f_L = \frac{0.63}{T} = 0.63\,\text{GHz}, \quad f_H = \frac{1.89}{T} = 1.89\,\text{GHz}, \quad B = \frac{1.26}{T} = 1.26\,\text{GHz} \quad (8.4)$$

8.4. SURFACE DISCRETIZATION AND FEED MODEL

The length of the antenna shown in Fig. 8.1 is approximately 6″. The structure is created using Matlab PDE toolbox. It has 550 triangles and 767 edge elements. The slot shape is described by an empirical formula [1]

$$\frac{\cos(l\pi)(1-\cos(l\pi))}{4} \quad (8.5)$$

where l the length along the slot. To model the antenna shown in Fig. 8.1 using the Matlab PDE toolbox, we first create a rectangular plate. Two raw slotlike polygons are then drawn inside the rectangle (left and right slots) using `polygon tool`. They follow Eq. (8.5). To precisely adjust slot dimensions, we open the corresponding m-file in the Matlab editor and change the vertex coordinates. To export the mesh to the main work space, we select the `Export Mesh` option from the `Mesh` menu.

After the mesh is initialized, the array of triangles, t, provides, in its fourth column, the domain number. In the present example all triangles of the plate except the slots belong to domain 1; triangles within the left slot belong to domain 2, and triangles within the right slot belong to domain 3. To "cut" those two slots, all triangles having a domain number other than 1 should be omitted. The result is demonstrated in Fig. 8.4a. The structure has 550 triangles and 767 edge elements. The corresponding mesh file `antenna01.m` is saved in the subdirectory `mesh`.

The enlarged feed area is shown in Fig. 8.4b and c. The feeding edge is indicated by a black bar. This is again the inner edge closest to the origin. For the slot antenna, the size of the feed edge is rather a free parameter due to the specific geometry of the slot. We therefore consider two different feed models shown in Fig. 8.4b and c. In Fig. 8.4b the feed edge has a length of 1 mm, whereas in Fig. 8.4c the edge length is 4 mm. We will see later that the antenna

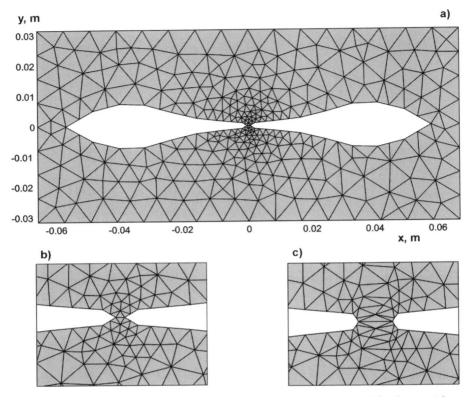

Figure 8.4. (a) Surface mesh created using PDE toolbox; (b, c) two tested feed geometries.

parameters are almost unaffected by the size of the feeding edge. Alternatively, the mesh for the slot antenna may be created using the Matlab function delaunay (see Chapter 7 for examples).

8.5. FREQUENCY LOOP

Scripts rwg1.m and rwg2.m are directly adopted from Chapter 7. They are not affected by the frequency loop and should be run only once. For the feed model in Fig. 8.4b, we use antenna01.mat as an input to rwg1.m. For the feed model in Fig. 8.4c, the file antenna02.mat should be used.

In the present example, we calculate the frequency response over the band from 12.5 MHz to 6.25 GHz with a sampling interval of 12.5 MHz. The number of sampling frequency points is 500. Such a large frequency spectrum is mostly chosen for the demonstration purposes. The reason is that possibly large time domain can be covered that contains both the radiated and received (delayed) signal. Running the script rwg31.m at these conditions requires 2.5 hours.

A moderate (about 15%) increase in speed may be obtained using the Matlab compiler. To do so, the line function []=rwg31; should be added at the beginning of the script. Then the file is compiled using the command mcc -m rwg31.m and executed as a stand-alone DOS application.

Sampling in the frequency domain is not equivalent to sampling in the time domain, which is discussed in many textbooks devoted to signals and systems [14,15]. In general, we should answer two questions: (1) how large the sampling interval in the frequency domain should be, and (2) what frequency band should be covered. The corresponding discussion is pursued in Sections 8.12 and 8.13. Already precalculated results for the frequency loop are saved in the Matlab directory of Chapter 8.

8.6. SURFACE CURRENT DISTRIBUTION

Figure 8.5 shows the surface current distribution on the antenna surface at different frequencies. The gray scale extends from the minimum to the maximum current magnitude in each case. The feed model antenna01.mat (Fig. 8.4b) is used.

In general, the surface current is concentrated close to the slot border and has the highest values at the antenna feed. This is seen in Fig. 8.5a at a relatively low frequency of 500 MHz as well as in Fig. 8.5c at a higher frequency of 2 GHz. Figure 8.5b (1 GHz) is an extension to the rule. The feed current appears to be very small and the antenna itself starts to "shine." We will see later that the present antenna type has a peak of the input impedance at approximately 1 GHz.

The surface current distribution in Fig. 8.5 is obtained using the script rwg5single.m which operates at every single frequency. The frequency (within the bandwidth) is specified at the beginning of the script. It is sometimes more convenient to plot the square root of the normalized current magnitude because the square root is more sensitive to low current densities. Actually Fig. 8.5 uses this type of scaling instead of the linear scale.

8.7. ANTENNA INPUT IMPEDANCE

The input antenna impedance is calculated in the script rwg31.m and is plotted using sweeplot.m. Figure 8.6 shows the impedance behavior as a function of frequency for two different models of the antenna feed.

In both cases the impedance peak is seen at approximately 1 GHz. Otherwise, the input impedance is predominantly real and has a relatively small variation within the bandwidth 1–5 GHz. It follows from Fig. 8.6 that two different feed models have nearly the same values of the input impedance in the entire frequency domain. Thus we may conclude that the input impedance of the slot antenna is weakly sensitive to the width of the feeding edge.

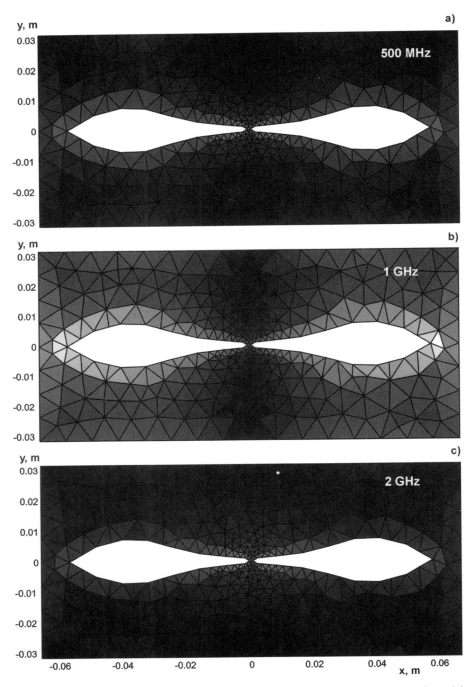

Figure 8.5. Surface current distribution (normalized magnitude) on the antenna surface: (a) 500 MHz; (b) 1 GHz; (c) 2 GHz. The color bar extends from the minimum to the maximum magnitude in each case.

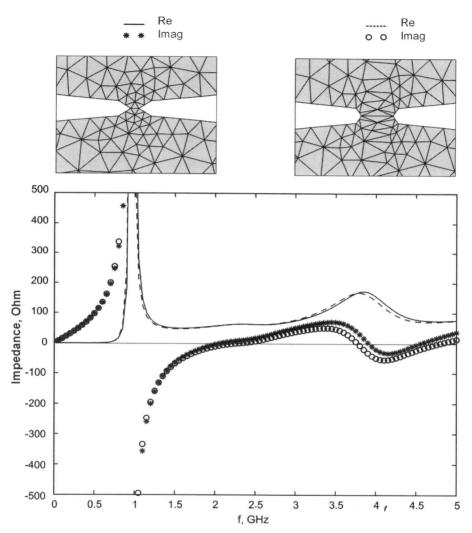

Figure 8.6. Input impedance as a function of frequency for two different feed models. Solid and dashed lines show the input resistance. The input reactance is shown by stars and circles.

Therefore only one feed model will be considered in the following, that is, `antenna01.mat`.

Compared to the broadband models considered in Chapter 7, the impedance behavior is superior to that of the bowtie antenna and comparable to the impedance variations of the spiral antenna. In practice, the input impedance of the slot antenna is affected by two attachment screws and other construction details.

The script `rwg31.m` simultaneously calculates the total power delivered to the antenna in the feed (radiated by an ideal antenna) using Eq. (4.16) of Chapter 4, that is,

$$P_{\text{feed}} = \frac{1}{2}\text{Re}(Z_A)|I|^2 = \frac{1}{2}\text{Re}(IV_A^*) \qquad (8.6)$$

where I is the total current through the feeding edge, Z_A is the antenna input impedance, and V_A is the feed voltage (1 V). For the transmitting antenna, V_A = 1 V. The impedance array `Impedance` and the power array `FeedPower` are filled at every frequency within the bandwidth. The power reaches its first maximum (resonance) of 9 mW at approximately 1.9 GHz. However, this maximum is weakly developed.

8.8. ANTENNA RADIATION INTENSITY, GAIN

For the far field, the script `efield2single.m` calculates the 3D radiation pattern at a single frequency within the bandwidth. Figure 8.7 shows the radiation intensity distribution at four frequencies: 500 MHz, 1 GHz, 2 GHz, and 4 GHz. The gray scale extends from minimum to maximum radiation intensity in each case. The antenna behaves like an omnidirectional (strictly speaking, bidirectional) radiator at frequencies below 4 GHz, where the maximum radiation intensity occurs in the axial direction (in the direction of the z-axis). However, at higher frequencies the direction of maximum radiation intensity may change.

The radiation patterns in Fig. 8.7 resemble those of the spiral antenna with the equivalent orientation in space (Fig. 7.16 of Chapter 7). We should, however, remember that the physical size of the slot antenna is nearly two times smaller. Therefore the frequency band for the purposes of comparison should be either extended to higher frequencies for the slot antenna or reduced to lower frequencies for the spiral antenna. In that sense the equivalent spiral antenna potentially has considerably smoother radiation patterns than the slot antenna of the present chapter. Furthermore, whereas the spiral antenna produces a nearly circular polarization, the slot antenna of this chapter is linearly polarized (its polarization is equivalent to that of the dipole perpendicular to the slot) in the direction of the main beam.

The polarization vector at a point can be calculated using the script `efield1single.m`. It should be noted that the present algorithm does not compute gain variations versus frequency. Instead, the algorithm of Chapter 7 can be used. For the present antenna, the logarithmic gain between 4 and 7 dB is observed in the frequency range 50 MHz to 6.25 GHz for both models of the antenna feed. The antenna gain is not needed for the transfer function, but it is important for the Friis transmission formula.

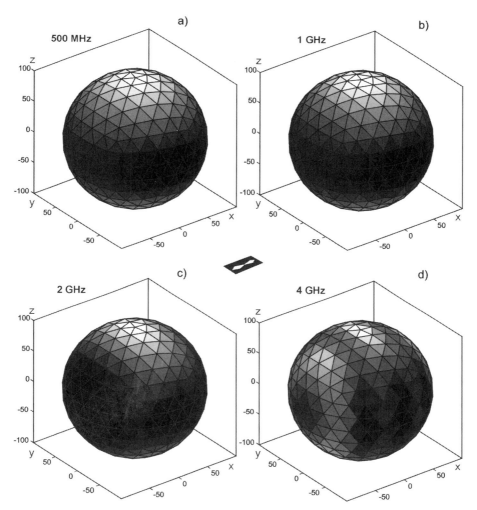

Figure 8.7. Radiation intensity distribution over the sphere surface with the radius of 100 m at four different frequencies.

8.9. DIRECTIVITY PATTERNS

As in Chapter 7 the script efield3single.m calculates the directivity patterns at a single frequency. The planar slot antenna is located in the xy-plane (see Fig. 8.4), with the slot direction being the x-direction and the axial direction being the z-direction. Therefore the most interesting for us are the xz-plane (the H-plane for the slot antenna) and the yz-plane (the E-plane). Figure 8.8 shows the radiation patterns at three different frequencies

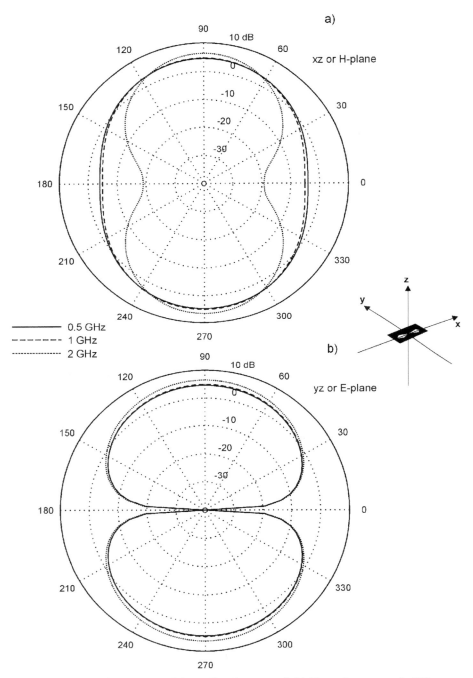

Figure 8.8. Directivity patterns of the radiated antenna field. The reference angle 90° corresponds to the z-axis.

(500 MHz, 1 GHz, and 2 GHz). As long as the frequency does not exceed 4 GHz, the gain in these planes coincides with the 3D maximum antenna gain.

The antenna directivity in the xz-plane shows large variations when frequency exceeds 2 GHz. The half-power beamwidth decreases from nearly 85° at 0.5 GHz to 60° at 2 GHz. In the yz-plane, the half-power beamwidth is kept nearly the same (about 85°) when frequency does not exceed 2 GHz. Therefore the present antenna cannot be considered as omnidirectional.

It is worth noting that the directivity pattern of the slot antenna in the yz-plane in Fig. 8.8b is essentially equivalent to the dipole radiation pattern in the yz-plane (Fig. 4.11a of Chapter 4). However, the dipole is oriented along the y-axis, whereas the slot axis is the x-axis.

The null radiation intensity appears on the y-axis for reason of symmetry. The condition of zero radiation in the y-direction holds for the symmetric slot antenna of an arbitrary shape.

When the frequency exceeds 4 GHz the directivity patterns show large variations. The main beam in the axial direction is destroyed. This was already seen in Fig. 8.7d.

8.10. ANTENNA-TO-FREE-SPACE TRANSFER FUNCTION

The antenna transfer function can be defined at a point as the radiated electric field at that point at different frequencies, under the assumption that the input voltage is kept at 1 V. To calculate the transfer function, the frequency loop is introduced into the script efield1.m as shown in Section 8.2. The resulting script is the script rwg32.m, and we should run it after rwg31.m. By default, the script outputs the electric field vector E(1:3,1:NumberOf-Steps) (both real and imaginary parts) at a number of sample frequencies into the binary file radiatedfield.mat.

Figure 8.9 shows the script output in the frequency range 12.5 MHz to 6.25 GHz. Included is the magnitude of the electric field as a function of frequency at the distance of 1 m from the antenna in the axial direction (the z-direction). The dominant y-component of the field, E_y, is shown. E_x is also observed, and it can be on the order of several percent of E_y at high frequencies. E_z is negligibly small in the entire frequency range.

Note in Fig. 8.9a that the present antenna acts like a band-pass filter with the center frequency of approximately 2 GHz. The phase in Fig. 8.9b was plotted using the Matlab command

```
plot(f,unwrap(angle(E(2,:))))
```

which returns the phase in radians, with an effort made to keep it continuous over the π-borders. It is also worth noting that the transfer function gives no indication of a minimum at 1 GHz (see Fig. 8.6), where the input antenna impedance becomes infinite. The reason is that we actually calculate the

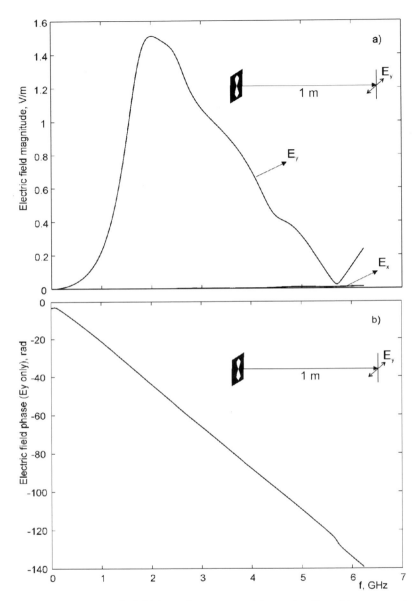

Figure 8.9. Radiated electric field (x- and y-components) as a function of frequency at a distance of 1 m in the axial direction from the transmitting antenna: (a) Magnitude; (b) phase of the y-component.

transfer function from the ideal voltage source in the antenna feed to a free space. If a fixed-impedance transmitter were introduced, then the result would change (see Section 8.12).

The antenna-to-free-space transfer function can be used to find the radiated pulse at an arbitrary spatial point. The method developed in Sections 8.12 and 8.13 can be applied for this purpose, without considerable modifications. The only difference is that the transfer function $T(n)$ in Eqs. (8.9) and (8.13), should be defined based on the array E(2,:) from the file radiatedfield.mat instead of the array OutputVoltage from the file receivedfield.mat. Both arrays have the same dimensions. The radiated pulse form in space is, however, auxiliary and not as important as the temporal pulse form received by the second antenna (Section 8.13).

We do not calculate the radiated pulse forms in this section since the corresponding analysis is very similar to the analysis performed in Sections 8.12 to 8.14 below for the receiving antenna. This task is addressed in a few problems at the end of the chapter.

A very important characteristic of the radiated pulse is its similarity in shape in different radiation directions. To quantitatively measure how similar the radiated electric field is in different directions, a "fidelity" parameter can be defined. The pulse *fidelity* is the maximum cross-correlation of the normalized radiated pulse p and a sample pulse. Mathematically [10]

$$\text{Fidelity} = \max_\tau \left(\int_{-\infty}^{\infty} p(t) p_{\text{sample}}(t+\tau) dt \right) \Big/ \sqrt{\int_{-\infty}^{\infty} p^2(t) dt \int_{-\infty}^{\infty} p_{\text{sample}}^2 dt} \qquad (8.7)$$

The fidelity value 1 means that two pulses are identical in shape. The value −1 means that the test pulse is an inverted replica of the sample pulse.

8.11. ANTENNA-TO-ANTENNA TRANSFER FUNCTION

After the radiated electric field at 1 m is found, we may place a second (identical or not) antenna at that point. The second antenna is the receiving antenna. In order to find the received voltage, the second antenna should be treated as a scatterer and the corresponding scattering problem should be solved. The script rwg33.m, which calculates antenna scattering, is a direct modification of the scripts rwg3.m and rwg4.m from Chapter 2. The two major changes include the frequency loop and the calculation of the received voltage. The incident field E at each frequency is obtained from the input file radiatedfield.mat. The wave vector (vector kv in script rwg33.m) is directed along the line connecting the antenna centers. In this chapter we investigate the case of two identical antennas.

Let us assume that the feeding edge has the number Index (e.g., Index =26 for the slot antenna). The total current through the edge is obtained as the product of the surface current and the edge length. The received voltage is the product of that current and the antenna impedance that is already calculated and saved in binary file current.mat. Thus the received voltage is obtained in the form

```
FeedCurReceived   =I(Index)*EdgeLength(Index);
FeedVolReceived   =FeedCurReceived*Impedance(FF);
OutputVoltage(FF)=FeedVolReceived;
```

where index FF corresponds to the discrete frequency domain. The output of the script is the array OutputVoltage(1:F) which contains the received voltage (complex number) at the discrete frequency points.

Furthermore the script rwg33.m outputs the total received power. The power can be calculated using two different methods: the direct method given by Eq. (4.23) of Chapter 4 (assuming the conjugate-matched load); and the Friis transmission formula given by Eq. (4.18) of Chapter 4, respectively. In Fig. 8.10 these two quantities are shown by solid and dashed lines, respectively. The coupled system transmitting-receiving antenna again acts very much like a band-pass filter with the center frequency at 2 GHz. Note that in order to

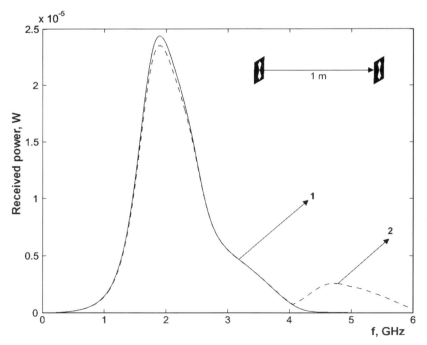

Figure 8.10. Total power received by the second antenna as a function of frequency (power spectrum of the antenna-to-antenna transfer function). (1) Direct power calculation, Eq. (4.23), (2) Friis transmission formula.

calculate the power using the Friis formula, we need to know the antenna gain as a function of frequency. The corresponding calculations were performed using the code of Chapter 7.

The agreement between two power curves in Fig. 8.9 is quite satisfactory except for the frequency domain above 4 GHz. In this domain the direction of maximum radiation is moved out of the antenna axis, and Eq. (4.18) of Chapter 4 predicts inadequate (too high) values of the power.

A note of caution should therefore be made with regard to the Friis transmission formula at high frequencies. If the direction of maximum radiation does not coincide with the direction to the receiving antenna, the gain in the direction to the receiving antenna should be substituted into Eq. (4.18) of Chapter 4 instead of the maximum antenna gain. Since the gain changes rapidly at high frequencies, accurate gain tracking may be difficult. The radiation intensity distribution, similar to that shown in Fig. 8.7, must be evaluated at each discrete frequency.

Once the output voltage of the second antenna is known as a function of frequency, we are able to predict the received voltage pulse for an arbitrary input voltage. This is the major point of the frequency domain analysis. The following sections explain how to obtain the received voltage using the discrete Fourier transform. The received voltage will be calculated in the script rwg34.m.

8.12. DISCRETE FOURIER TRANSFORM

As specified in Section 8.5, the discrete frequency domain covers the frequency band 12.5 MHz – 6.25 GHz with a sampling interval of 12.5 MHz. The frequency points are

$$f(n) = \Delta f n; \quad n = 1, \ldots, F; \quad \Delta f = 12.5 \times 10^6; \quad F = 500 \tag{8.8}$$

The antenna-to-antenna transfer function (array OutputVoltage) is evaluated at these frequency points and is denoted here by $T(n)$. To apply the discrete Fourier transform (DFT), the transfer function must be defined at the corresponding negative frequencies as well, i.e. at $n = -F, \ldots, -1$. There is no need to perform additional calculations. The transfer function value at a negative frequency, $f(-n)$, is the complex conjugate of the corresponding value at the positive frequency, $f(n)$.

The next step is to convert the frequency sequence from $-F$ to F to the sequence from 0 to $2F$, that is the convenient practice for the DFT [16, p. 153]. This is done using the formula

$$T(n) = T^*(N + 1 - n), \quad n = F + 1, \ldots, N, \quad N = 2F \tag{8.9}$$

The Matlab code

```
for n=F+1:N
    OutputVoltage(n)=conj(OutputVoltage(N+1-n));
end
```

in the script `rwg34.m` performs the same operation and is equivalent to Eq. (8.9). The transfer function is now defined at the discrete frequency points $n = 1, \ldots, N$. Since we do not have a dc component, neither in the pulse spectrum nor in the transfer function, the value at $n = 0$ is always equal to zero and can be ignored.

The DFT of a voltage pulse $p(t)$ can be written in the following form [14–16]:

$$P(n) = \sum_{k=0}^{N-1} p(k) \exp\left(\frac{-j2\pi kn}{N}\right) \qquad (8.10)$$

Here $P(n)$ are N values of the frequency spectrum of a pulse at frequencies $f(n)$; $p(k)$ are N values of the time domain pulse form sampled at times $t - Tk$. The discrete Fourier transform implies that the sampling time interval, T, equals

$$T \equiv \frac{1}{\Delta f N} = \frac{1}{2 \max(f(n))} \qquad (8.11)$$

Substitution of the corresponding values from Eq. (8.8) shows that $T = 0.08$ ns if the upper frequency is 6.25 GHz.

The inverse DFT is given by [14–16]

$$p(k) = \frac{1}{N} \sum_{n=0}^{N-1} P(n) \exp\left(\frac{+j2\pi kn}{N}\right) \qquad (8.12)$$

The standard DFT (Matlab functions `fft` and `ifft`) employs N time samples and N frequency samples. Sometimes, the number of time samples is not equal to the number of frequency samples [14,16]. The script `rwg34.m` allows varying the number of time samples and the size of the time window in order to simultaneously observe the transmitted pulse and a time-delayed received pulse. Note that for a pulse with zero dc content, the summation in Eqs. (8.10) and (8.12) can be done from 1 to N instead of 0 to $N - 1$. The script `rwg34.m` uses directly the DFT method described by Eqs. (8.10) and (8.12).

8.13. RECEIVED VOLTAGE PULSE

With the help of the previous section, the temporal form of the received voltage pulse, $p_R(t)$, is obtained using the spectrum multiplication and the inverse DFT

$$p_R(k) = \frac{1}{N}\sum_{n=0}^{N-1} P_R(n)\exp(+j2\pi kn/N); \quad P_R(n) = P(n)T(n) \qquad (8.13)$$

Here $P(n)$ is the frequency spectrum of the transmitted pulse given by Eq. (8.10); $T(n)$ is the antenna-to-antenna transfer function from Eq. (8.9). The spectrum multiplication is performed in the script rwg34.m. Finally, Eqs. (8.12) and (8.13) output two temporal pulse forms: one of the original transmitted pulse and another of the received pulse. Figure 8.11 shows the script output for three Gaussian pulses (Section 8.3) of different durations.

The pulse durations are 2, 1, and 0.25 ns. The solid line is the voltage pulse across the feed of the first antenna; the dotted line is the received voltage across the feed of the second antenna. The second (receiving) antenna is placed 1 m apart from the first antenna (Fig. 8.10).

If the input pulse duration is 2 ns, then the received pulse is totally distorted (Fig. 8.11a). The reason for such a distortion is the large discrepancy between the pulse center frequency and the pass band of the antenna-to-antenna filter shown in Fig. 8.10. According to Eq. (8.4), the pulse center frequency in that case is 0.6 GHz and is considerably below the pass band of the band-pass filter in Fig. 8.10. A similar situation occurs when the center frequency is too high (beyond the pass band), meaning that the pulse length is too short. Figure 8.11c gives the results for the pulse of 0.25 ns duration with the center frequency at 4.8 GHz. The received pulse appeared longer in duration than expected, having relatively small power.

The best transmission results are obtained when the pulse center frequency and the center frequency of the antenna-to-antenna transfer function are close to each other. Figure 8.11b shows the received voltage pulse if the transmitted pulse has the duration of 1 ns (center frequency is 1.2 GHz). The received pulse has nearly the same duration and the highest peak voltage compared to other cases. Similar results are obtained for the pulse of 0.5 ns duration (center frequency is 2.4 GHz). It is interesting that the received pulse at 1 ns resembles the second derivative of the initial pulse. The shape of the second derivative is shown in Fig. 8.11b within a box. The present results are still an idealized approach in the sense that we do not take into account the coupling of the transmitting and receiving antennas to a load. This question is addressed in the next section.

Note that the results for the 0.25 ns long pulse may not be very accurate when the maximum sampling frequency is 6.25 GHz. A better approach is to double this value.

8.14. IMPEDANCE MISMATCH

A more realistic model of the transmitting antenna is shown in Fig. 8.12a. The antenna, represented by an equivalent input impedance Z_A, is connected to a generator with the characteristic impedance Z_G. The incident voltage pulse is

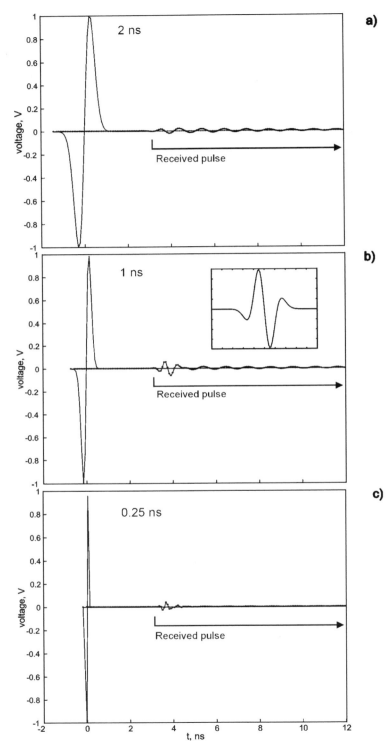

Figure 8.11. Transmitted (solid line) and received (dotted line) voltage pulses in the antenna feed. (a) Pulse duration is 2 ns; (b) 1 ns; (c) 0.25 ns.

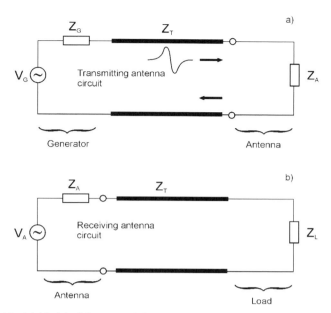

Figure 8.12. (a) Model of the transmitting antenna; (b) model of the receiving antenna.

sent by a pulse generator to the antenna through the transmission line of the same impedance $Z_T = Z_G$. Since there is usually a mismatch between Z_A and Z_G, part of the pulse will be reflected back toward the generator.

Assuming an infinitely long transmission line, we obtain the *reflection coefficient*, Γ, of a harmonic wave of frequency ω in the form (see [17, pp. 398–399] or [18, p. 502])

$$p^-(\omega) = \frac{Z_A - Z_T}{Z_A + Z_T} p^+(\omega) = \Gamma(\omega) p^+(\omega) \qquad (8.14)$$

where superscripts ± denote the incident and reflected wave, respectively. The transmission line typically has the frequency-independent real impedance of $Z_T = 50\,\Omega$. The antenna impedance is certainly different from that value (see Fig. 8.6). Therefore a reflection takes place across the antenna feed. The total voltage across the feed is the sum of the transmitted and reflected signals:

$$p(\omega) = p^+(\omega) + p^-(\omega) = \left[\frac{2Z_A}{Z_A + Z_T} \right] p^+(\omega) \qquad (8.15)$$

The quantity in square brackets is the *transmission coefficient* of the antenna. We assume that the incident voltage, $p^+(\omega)$, has the form of the Gaussian monopulse (8.1). In order to account for the distortion of the incident pulse at the antenna feed, the transmission coefficient (8.15) should be inserted into Eq. (8.13).

Another addition implies the receiving circuit shown schematically in Fig. 8.12b. A convenient way to model the receiving antenna is using an ideal voltage source of strength V_A in series with the impedance Z_A (Thévenin equivalent) [19]. In our case, V_A is the voltage across the feed of the receiving antenna calculated in Section 8.11; Z_A is the input antenna impedance calculated in Section 8.7. The receiving antenna is also connected to an infinite transmission line, whose impedance, Z_T, is equal to the load impedance, Z_L. According to the voltage division principle, the voltage delivered to the transmission line (to the load) is equal to

$$\left[\frac{Z_T}{Z_A + Z_T}\right] V_A(\omega) \tag{8.16}$$

To summarize these two contributions, a factor

$$\left[\frac{2Z_A}{Z_A + Z_T}\right]\left[\frac{Z_T}{Z_A + Z_T}\right] \tag{8.17}$$

where $Z_T = 50\,\Omega$, should be inserted into Eq. (8.13). Thus the corrected version of Eq. (8.13) takes the form

$$p_R(k) = \frac{1}{N}\sum_{n=0}^{N-1} P_R(n)\exp\left(\frac{+j2\pi kn}{N}\right), \quad P_R(n) = \frac{2Z_A(n)Z_T}{(Z_A(n)+Z_T)^2} P(n)T(n) \tag{8.18}$$

or, in terms of the corresponding Matlab code,

```
SPECTRUM=OutputVoltage.*FDIRECT.*...
(2*Impedance./(Impedance+50)).*(50./(Impedance+50));
```

Equation (8.18) assumes that a Gaussian pulse, $p(t)$, from a pulse generator is sent to the transmitting antenna through a 50 Ω transmission line. In turn the received pulse, $p_R(t)$, is now the voltage across the load connected to the antenna through the infinite transmission line with the characteristic impedance of 50 Ω. These assumptions would closely correspond to a real communication system if we could afford ideal impedance matching between various system components.

8.15. VOLTAGE PULSE AT A LOAD

The script `rwg34.m` also calculates the received pulse, $p_R(t)$, assuming the 50 Ω transmission lines for the receiving and transmitting antennas. The more realistic model of Eq. (8.18) is used. Figure 8.13 shows the script output

VOLTAGE PULSE AT A LOAD 217

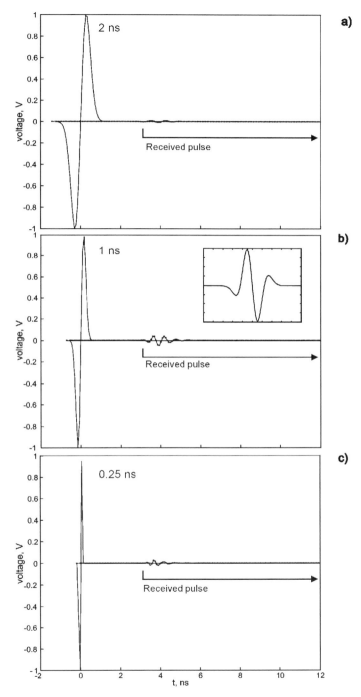

Figure 8.13. Transmitted (solid line) pulse at the generator and received (dotted line) voltage pulse at the load. (a) Pulse duration is 2 ns; (b) 1 ns; (c) 0.25 ns.

for three Gaussian pulses with durations 2, 1, and 0.25 ns (50 Ω matching impedance).

The solid line is the transmitted voltage pulse at the generator, whereas the dotted line indicates the received voltage across the load. Both voltages are plotted using the *same* scale. The second receiving antenna is again placed 1 m apart from the first antenna (Fig. 8.10).

It is useful to compare Figs. 8.11 and 8.13 in order to see the differences produced by the impedance mismatch between the antenna and the transmission line. For a 2 ns long pulse, the received pulse in Fig. 8.13a is almost zero, whereas Fig. 8.11a predicts a reasonably high received voltage. Thus the received voltage is expected to be smaller than in the ideal case.

Fortunately, in the most important case of the 1 ns long pulse, the impedance mismatch has almost no effect on the received voltage. If we compare Fig. 8.11b and Fig. 8.13b, the only difference we find is a slight change in the shape of the received pulse. The received pulse is no longer the second derivative of the incident voltage depicted in a box in Fig. 8.13b. At higher pulse durations (Figs. 8.11c and 8.13c), the impedance mismatch leads to a drop in the pulse magnitude but does not change the pulse form considerably.

The present investigation shows that the slot antenna introduced in this chapter is well suited for the transmission of 1 ns long pulses. For shorter, and especially longer, pulses its performance is rather unsatisfactory. The further optimization of the antenna geometry seems to be possible and desirable. Also the antenna performance at different elevation angles should be investigated to complete the study.

The model of the infinitely long transmission line does not take into account the multiple reflection of the pulse in the transmitting circuit. As a result of these reflections, a delayed pulse replica can be radiated one or more times. This replica, if not properly damped, may destroy results shown in Fig. 8.13.

8.16. CONCLUSIONS

The time domain analysis by the MoM method presented in this chapter requires rather lengthy calculations. However, it can be done routinely, using the well-established MoM solver in the frequency domain. The problems at the end of the chapter extend the method to other broadband antenna types. That is to say, the reader can check the performance of the spiral antenna and the bowtie antenna for pulse transmission.

Another possible time domain analysis consists in formulating the original integral equations not in the frequency domain but in the time domain [20]. In this method we do not even need to solve the matrix equations because the solution can be obtained using a time domain propagator (the so-called marching-on-in-time method) [20]. The method, however, suffers from instabilities and requires some special computational tools [21].

REFERENCES

1. M. A. Barnes. *Ultra-wideband Magnetic Antenna*. US Patent 6,091,374 of July 18, 2000.
2. Y. Nishioka, O. Maeshima, T. Uno, and S. Adachi. FDTD analysis of resistor-loaded bow-tie antennas covered with ferrite coated conducting cavity for subsurface radar. *IEEE Trans. Antennas and Propagation*, 47 (6): 970–977, 1999.
3. E. S. Eide. Ultra-wideband transmit/receive antenna pair for ground penetrating radar. *IEE Proc. Microwave Antennas Propagation*, 147 (3): 231–235, 2000.
4. D. A. Kolokotronis, Y. Huang, and J. T. Zhang. Design of TEM horn antennas for impulse radar. In *High Frequency Postgraduate Student Colloquium*, 1999, University of Leeds, pp. 120–126.
5. M. Z. Win and R. A. Scholtz. Impulse radio: How it works. *IEEE Communication Letters*, 2: 36–38, 1998.
6. J. T. Conroy, I. L. LoCicero, and D. R. Ucci. Communication techniques using monopulse waveforms. In *MILCOM 1999 IEEE Military Communications Conference Proc.* IEEE Press, Piscataway, NJ, 1999, pp. 1181–1185.
7. S. C. Hagness, A. Taflove, and J. E. Bridges. Wideband ultralow reverberation antenna for biological testing. *Electronics Letters*, 33 (19): 1594–1595, 1997.
8. S. C. Hagness, A. Taflove, and J. E. Bridges. Three-dimensional FDTD analysis of pulsed microwave confocal system for breast cancer detection: design of an antenna array element. *IEEE Trans. Antennas and Propagation*, 47 (5): 783–791, 1999.
9. K. L. Schlager, G. S. Smith, and J. G. Maloney. Optimization of bow-tie antennas for pulse radiation. *IEEE Trans. Antennas and Propagation*, 42 (7): 975–982, 1994.
10. T. P. Montoya and G. S. Smith. A study of pulse radiation from several broad-band loaded monopoles. *IEEE Trans. Antennas and Propagation*, 44 (8): 1172–1182, 1996.
11. A. Taflove and S. Hagness. *Computational Electrondynamics: The Finite-Difference Time-Domain Method*, 2nd ed. Artech House, Boston, 2000.
12. K. S. Kunz and R. J. Luebbers. *The Finite Difference Time Domain Method for Electromagnetics*. CRC Press, Boca Raton, 1993, ch. 3.
13. K. L. Virga and R. Y. Rahmat-Samii. Efficient wide-band evaluation of mobile communications antennas using [Z] or [Y] matrix interpolation with the method of moments. *IEEE Trans. Antennas and Propagation*, 47 (1): 65–76, 1999.
14. S. Haykin and B. V. Veen. *Signals and Systems*. Wiley, New York, 1998.
15. B. P. Lathi. *Linear Systems and Signals*. Berkeley-Cambridge Press, Carmichael, CA, 1992.
16. K. Steiglitz. *A Digital Signal Processing Primer*. Addison-Wesley, Menlo Park, CA, 1996.
17. K. R. Demarest. *Engineering Electromagnetics*. Prentice Hall, Upper Saddle River, NJ, 1998.
18. J. D. Kraus. *Electromagnetics*, 4th ed. McGraw Hill, New York, 1992.
19. D. M. Pozar. *Microwave and RF Design of Wireless Systems*. Wiley, New York, 2001.

20. A. J. Poggio and E. K. Miller. Integral equation solutions of three-dimensional scattering problems. In R. Mittra, ed., *Computer Techniques for Electromagnetics*, 2nd ed. Pergamon, Oxford, 1973, pp. 159–264.
21. G. Marana and A. Monorchio. Experimental validation of a stable method of moments procedure in time domain for the scattering from arbitrarily shaped conducting bodies. *Radio Science*, 35 (4): 923–931, 2000.

PROBLEMS

8.1. Using the script bowtie.m from subdirectory mesh, create a structure for a bowtie antenna of total length 13.4 cm[2], a flare angle of 90°, and a feeding edge length of 5 mm. Plot the antenna-to-free-space transfer function (amplitude and phase of the dominant E-component) 1 m apart on the antenna axis (the z-axis) in the range 25 MHz to 6.0 GHz. The number of sampling frequency points is 50.

8.2. Using the script spiralplane.m from subdirectory mesh, create a structure for the Archimedean spiral antenna of total size 13.4 cm[3] with 5 turns and a strip width of 10 mm. Plot the antenna-to-free-space transfer function (amplitude and phase of two dominant E-components) 1 m apart on the antenna's axis (the z-axis) in the range 25 MHz to 6.0 GHz. The number of sampling frequency points is 50.

8.3. Using the structure antenna01.mat for the slot antenna, plot the antenna-to-free-space transfer function (amplitude and phase of the dominant E-component) at 1 m from the antenna center, in the xz-plane, at elevation angles 15° and 45°, respectively. The frequency range is 100 MHz to 6.0 GHz. The number of sampling frequency points is 50.

8.4. Repeat Problem 8.3 for a bowtie antenna of total length 13.4 cm (the size of the slot antenna from the present chapter), a flare angle of 90°, and a feeding edge length of 5 mm.

8.5. The voltage across the feed of the transmitting slot antenna shown in Fig. 8.4*a* (antenna01.mat) is a Gaussian pulse of 1 ns duration. Determine and plot the radiated pulse (dominant component of the electric field) on the antenna axis (the z-axis) at $z = 0.5$ m.

8.6. The voltage across the feed of a transmitting bowtie antenna of total length 13.4 cm, a flare angle of 90°, and a feeding edge length of 5 mm is a Gaussian pulse of 1 ns duration. Determine and plot the radiated pulse (dominant component of the electric field) on the antenna axis (the z-axis) at $z = 0.5$ m.

---

[2] 13.4 cm is the size of the slot antenna of the present chapter in the direction of the slot (see Fig. 8.4).
[3] Ibid.

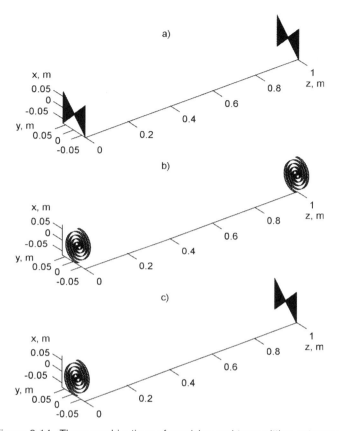

Figure 8.14. Three combinations of receiving and transmitting antennas.

8.7. The voltage across the feed of a transmitting Archimedean spiral antenna of total size 13.4 cm with 5 turns and the strip width 10 mm is a Gaussian pulse of 1 ns duration. Determine and plot the radiated pulse (two dominant components of the electric field) on the antenna's axis (the z-axis) at $z = 0.5$ m.

8.8.* The voltage across the feed of a transmitting bowtie antenna of total length 13.4 cm, a flare angle of 90°, and a feeding edge length of 5 mm is a Gaussian pulse of 1 ns duration. Determine and plot the voltage pulse received by a second identical bowtie antenna (voltage across the feed). The receiving antenna is located at 1 m, face to face to the transmitting antenna (Fig. 8.14a).

8.9.* The voltage across the feed of a transmitting Archimedean spiral antenna of total size 13.4 cm with 5 turns and the strip width 10 mm is a Gaussian pulse of 1 ns duration. Determine and plot the voltage pulse received by a second identical spiral antenna (voltage across the feed). The receiving antenna is located at 1 m, face to face to the transmitting antenna (Fig. 8.14*b*).

8.10.** The voltage across the feed of a transmitting Archimedean spiral antenna of total size 13.4 cm with 5 turns and the strip width 10 mm is a Gaussian pulse of 1 ns duration. Determine and plot the voltage pulse received by a second bowtie antenna of total length 13.4 cm, a flare angle of 90°, and a feeding edge length of 5 mm. The receiving antenna is located at 1 m, face to face to the transmitting antenna (Fig. 8.14*c*).

9

ANTENNA LOADING: LUMPED ELEMENTS

9.1. Introduction
9.2. Code Sequence
9.3. Lumped Resistor, Inductor, and Capacitor
9.4. Test
9.5. Effects of Resistive and Capacitive Loading
9.6. Conclusions
   References
   Problems

9.1. INTRODUCTION

The transient response of perfectly conducting cylindrical electric dipoles and monopoles is characterized by a number of reflections from the ends of the antenna (typically from the free end), resulting in an erratic response long after the excitation pulse is ended. This is why the standard dipole/monopole cannot be used for the broadband antenna design (for UWB purposes).

A classical approach to prevent the reflection from the end of a dipole or monopole antenna is to introduce a variable internal resistance per unit length. This is the so-called Wu and King model [1,2]. Although the antenna is purely reflectionless at only one particular frequency, the frequency dependence is weak, resulting in a current distribution that is very nearly *independent* of frequency [3].

Another approach to prevent the reflection from the end of the dipole or monopole, keeping the ohmic losses negligible, is to use a variable capacitance per unit length. It originates with Hallén [4]. The same objective is achieved of nearly occluding the reflecting wave from the end of the monopole but by way of a reactive distributed impedance instead of the real one. The corresponding theory somewhat follows the Wu and King model. The reasoning behind such a replacement is that in eliminating the ohmic losses of pure resistive loading, one could increase the antenna's efficiency as a radiator. Capacitive loading was further investigated in [5,6]. A comprehensive review of different loading types is given in [7]. The NEC code for loaded antennas is discussed in [8].

In this very short chapter we study the lumped and distributed loading of the simple antenna types. The RWG edge elements are ideally suited for modeling the antenna loading. Actually we only need to change slightly the diagonal terms of the impedance matrix. The corresponding theory was developed in [9,10].

9.2. CODE SEQUENCE

The code sequence of Chapter 7 remains practically unchanged. The only script that is changed is rwg3.m. This script has a new block that introduces the lumped loading of a few RWG edge elements. Distributed loading can be simulated using a straightforward modification of this block.

9.3. LUMPED RESISTOR, INDUCTOR, AND CAPACITOR

Figure 9.1 illustrates the idea of the loaded edge (RWG edge element) for four different loading types with RWG edge elements [9]. One of them, which is the voltage feed, has been already investigated in the preceding chapters. Equally with the voltage feed, we can introduce a resistive, inductive, or capacitive load of an edge element.

To see how the loaded edge works we return to the voltage feed model discussed in Chapter 4. First, we treat a loaded edge n as if it were exactly the voltage source edge, with an unknown voltage in the feed V. Next, the unknown voltage is expressed in terms of the total current through the edge n. According to Eq. (4.7) this current is equal to $l_n I_n$, where l_n is the edge length. The corresponding expression is given by

$$V = z_n(l_n I_n) \tag{9.1}$$

where z_n is the lumped impedance (or resistance for a purely resistive load). Eq. (9.1) is none other than circuit Ohm's law for an impedance element z_n.

Next, the voltage drop (9.1) is incorporated into the moment equations according to Eq. (4.5) and moved to its left-hand side (to the impedance matrix). Since the source voltage and the load voltage must have the opposite

LUMPED RESISTOR, INDUCTOR, AND CAPACITOR 225

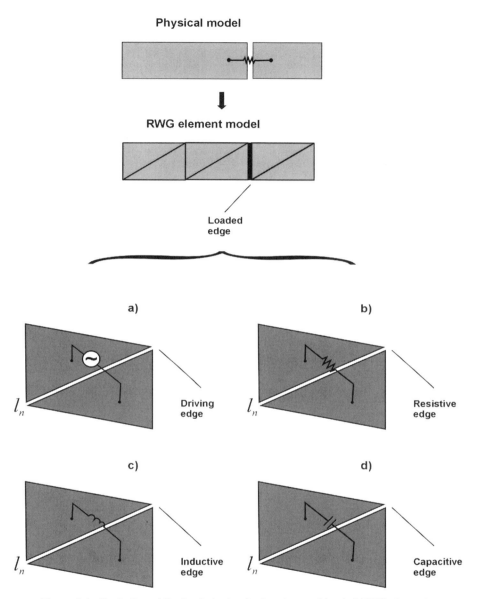

Figure 9.1. Illustration of the loaded edge for four types of loaded RWG elements.

polarities (see the passive and active reference configurations for an electric circuit), the sign of the impedance contribution must be the opposite of the feed voltage contribution (alternatively, we could use—V instead of V).

Thus, if the edge n is the only loaded edge, only one diagonal element of the impedance matrix Z has to be modified according to the following formulas (see also [9]):

$$\text{Resistance } R: \quad Z_{nn} \to Z_{nn} + (l_n)^2 R$$

$$\text{Capacitance } C: \quad Z_{nn} \to Z_{nn} + (l_n)^2 \frac{1}{j\omega C}$$

$$\text{Inductance } L: \quad Z_{nn} \to Z_{nn} + (l_n)^2 j\omega L$$

$$\text{Impedance } Z: \quad Z_{nn} \to Z_{nn} + (l_n)^2 Z \qquad (9.2)$$

The impedance contribution is programmed in the script rwg3.m. First, we identify positions of the lumped elements and the LCR (inductance/capacitance/resistance) values for every element. The "direction" of the lumped element has to be given in order to make clear exactly which edge is loaded. The code below illustrates these steps:

```
%IMPEDANCE ELEMENTS
    %Lumped impedance format
    % LoadPoint       Lumped element locations
    % LoadValue       Vector of L, C, and R
    % LoadDir         "Direction" of lumped element
LNumber=2;
LoadPoint(1:3,1)=[0  0.50 0]';
LoadValue(1:3,1)=[0  1e16 100]';  %LCR
LoadDir  (1:3,1)=[0  1 0]';
LoadPoint(1:3,2)=[0  -0.50 0]';
LoadValue(1:3,2)=[0  1e16 100]';  %LCR
LoadDir  (1:3,2)=[0  1 0]';
for k=1:LNumber
     DeltaZ(k)=j*omega*LoadValue(1,k)+...
     1/(j*omega*LoadValue(2,k))+LoadValue(3,k);
end
```

Next, a corresponding modification of the impedance matrix is made for some edge elements. These elements are identified in the following way: the corresponding edges are those closest to the load points and are perpendicular to the associated load "directions."

The concept of impedance loading is physically very clear for thin strips or wires, where we have only one RWG edge element per strip width. The loaded edge corresponds to a resistor, or a inductor, or a capacitor that breaks the thin strip or the thin wire. Many practical realizations of this design are discussed in [8].

At the same time we can formally introduce lumped loading for plates or other surfaces. It is, however, a difficult question of how to realize such a loading in practice. One way might be to cut narrow slots in the metal surface and "shorten" them by the lumped elements.

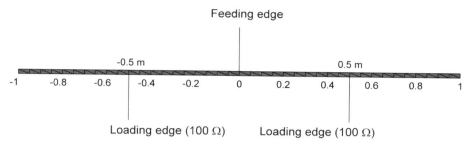

Figure 9.2. Feed and load positions for the 2 m long dipole.

Table 9.1. Radiation Characteristics of the Resistively Loaded Dipole 75 MHz

Model	Input Impedance, Ω	Feed Power, W	Radiated Power, W
Present calculation	$188 - j \times 14$	0.0026	0.0011
Equivalent wire monopole (SuperNEC) with 39 segments	$198 + j \times 0.1$	0.0025	0.0012

9.4. TEST

As a test example we consider the 2 m long dipole simulated by a strip with 80 triangles and a width of 0.02 m. This corresponds to a wire radius of 0.005 m. The structure is generated using the script strip.m in subdirectory mesh. Two loading edges are shown in Fig. 9.2. We investigate the resistive lumped loading using two resistors, 100 Ω each. The frequency is 75 MHz (half-wavelength dipole).

Table 9.1 compares the simulation results of the present chapter with the SuperNEC simulation results [11] for a loaded wire dipole of an equivalent radius. The agreement is quite satisfactory. Table 9.1 also indicates the difference between the radiated power of the lossy antenna and the power delivered to the antenna in the feed. The former value is considerably smaller than the latter. In the present case the ratio of two powers (*antenna efficiency*) is only about 50%. The feed power (at 1 V feed voltage) is calculated in the script rwg3.m, whereas the total radiated power (array TotalPower) is found in the script efield2.m.

9.5. EFFECTS OF RESISTIVE AND CAPACITIVE LOADING

First, we investigate the antenna performance in the frequency domain for the resistive loading. The same 2 m long strip is considered, with the same position of the loading elements. Parameters of the frequency loop are specified in the script `rwg3.m`. We choose the band from 25 to 500 MHz with totally 200 frequency steps. Since the strip only has 79 edge elements, the frequency loop does not take more than 1.5 minutes. The impedance parameters are obtained by running the script `sweeplot.m`. Note that this script also outputs the return loss of the antenna as a function of frequency.

Figure 9.3 compares three plots for the input impedance: the unloaded dipole (*a*), the dipole loaded with two 100 Ω resistors (*b*), and the dipole loaded with two 200 Ω resistors (*c*). Notice that the moderate resistive loading (Fig. 9.3*b*) still preserves the impedance peaks but makes the impedance behavior smoother compared to the unloaded dipole. The strong resistive loading may completely change the impedance curves (Fig. 9.3*c*). As can be seen, the position of the load plays a significant role for the dipole. In particular, the second resistance peak in Fig. 9.3 remains practically unchanged, since there is almost no current through the resistors at the corresponding frequency.

To obtain the reflectionless dipole, the resistive loading per unit length (*y* changes from –*h* to *h*) should be given by [1,7] (in Ω/m)

$$R = \frac{R_0}{1 - |y|/h} \qquad (9.3)$$

The value of R_0 (in Ω/m) is calculated separately [1]. The variable internal resistance can be implemented as a thin conductive tube of *variable* thickness, surrounding a nonconductive dielectric rod [7]. The dependency (9.2) can be approximated by a number of lumped resistors as well.

Next, the capacitive loading is studied. Figure 9.4 compares three plots for the input impedance of an unloaded dipole (*a*), a dipole loaded with two 10 pF capacitors (*b*), and a dipole loaded with two 1 pF capacitors (*c*). The position of the load corresponds to that in Fig. 9.2.

As shown in the figure, the capacitive loading makes the impedance curves smoother compared to the unloaded dipole. Furthermore the 10 pF capacitive loading shifts the impedance curves slightly to the right. In other words, the capacitive loading makes the antenna "shorter" than it physically is. On the other hand, an inductive loading can be shown to make the antenna "longer" than it is. This inviting feature of inductive/capacitive loading is widely used in practice.

When the loading capacitance is very small (impedance magnitude is very high),[1] the impedance behavior changes completely. This is shown in Fig. 9.4*c*.

9.6. CONCLUSIONS

In this chapter we briefly discussed a simple model of lumped impedance loading as adapted to RWG edge elements. Although the impedance loading

---

[1] It may be difficult to realize such small capacitance in practice.

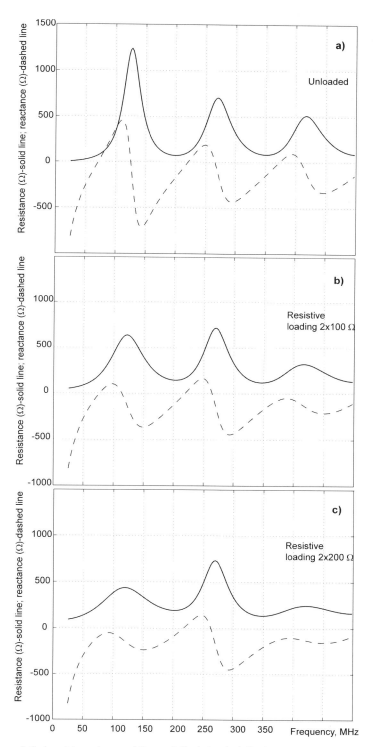

Figure 9.3. Input impedance of the resistively loaded dipole at different load values.

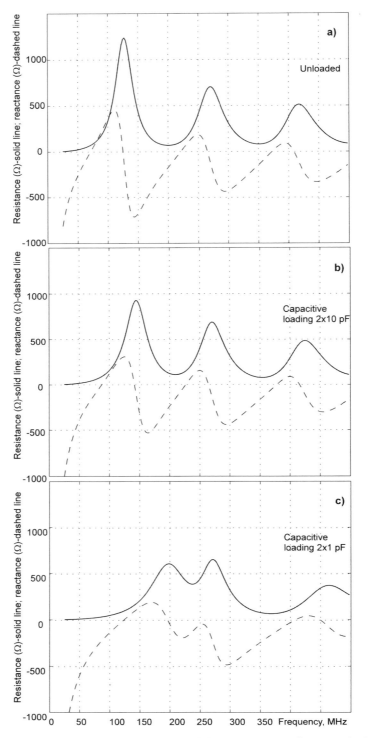

Figure 9.4. Input impedance of the capacitively loaded dipole at different load values.

is incorporated very simply into to the numerical model, its practical realization can be difficult [7,8], especially at higher frequencies [8]. Resistive or capacitive, or resistive/capacitive, loading is generally used to design broadband nonresonant dipoles [7]. Distributed loading can approximately be represented as a number of lumped elements with the adequate impedance per unit length (area).

REFERENCES

1. T. T. Wu and R. W. P. King. The cylindrical antenna with nonreflective resistive loading. *IEEE Trans. Antennas and Propagation*, 13: 369–373, 1965.
2. T. T. Wu and R. W. P. King. The cylindrical antenna with nonreflective resistive loading—Corrections. *IEEE Trans. Antennas and Propagation*, 13: 998, 1965.
3. G. C. Rose and R. S. Vickers. Transient response of resistively loaded cylindrical antennas. *Int. Journal of Electronics*, 36 (4): 479–486, 1974.
4. E. Hallén. *Electromagnetic Theory*. Wiley, New York, 1962, at paragraph 35.9 "Reflection-free antennas," pp. 501–504.
5. B. L. J. Rao, J. E. Harris, and W. E. Zimmerman. Broadband characteristics of cylindrical antennas with exponentially tapered capacitive loading. *IEEE Trans. Antennas and Propagation*, 17: 145–151, 1969.
6. M. Kanda. Time-domain sensors for radiated impulsive measurements. *IEEE Trans. Antennas and Propagation*, 31: 438–444, 1983.
7. T. P. Montoya and G. S. Smith. A study of pulse radiation from several broad-band loaded monopoles. *IEEE Trans. Antennas and Propagation*, 44 (8): 1172–1182, 1996.
8. B. D. Popović, M. B. Dragović, and A. R. Djordjević. *Analysis and Synthesis of Wire Antennas*. Wiley, New York, 1982.
9. D. Jiao and J.-M. Jin. Fast frequency-sweep analysis of RF coils for MRI. *IEEE Trans. Biomedical Engineering*, 46 (11): 1387–1390, 1999.
10. C. J. Leat, N. V. Shuley, and G. F. Stickley. Triangular-patch modeling of bowtie antennas: Validation against Brown and Woodward. *IEE Proc. Microwave Antennas Propagation*, 145 (6): 465–470, 1998.
11. *MoM Technical Reference Manual*. Poynting Software Ltd., 2001, 66 pp.

PROBLEMS

9.1. For a 2 m long dipole estimate the effect of inductive loading on the input impedance and the return loss. Use a frequency sweep of 75 to 500 MHz with a total of 200 frequency steps. The loaded elements are shown in Fig. 9.2. The inductance value is $0.1\,\mu$H per element.

9.2. Repeat Problem 9.1 if the inductance value changes to $1\,\mu$H.

232 ANTENNA LOADING: LUMPED ELEMENTS

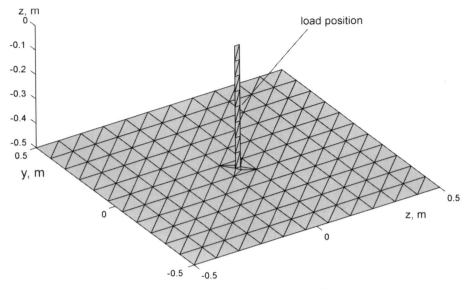

Figure 9.5. Monopole for Problem 9.5.

9.3. For a 2 m long dipole estimate the effect of combined inductive/capacitive loading on the input impedance and the return loss. Use a frequency sweep of 75 to 500 MHz with a total of 200 frequency steps. The position of the two loading elements is shown in Fig. 9.2. The capacitance value is 10 pF (for each element). The inductance value is $0.1\,\mu H$ (for each element). What is the antenna efficiency then?

9.4. For a 2 m long dipole estimate the effect of the combined inductive/resistive loading on the input impedance and the return loss. Use a frequency sweep of 75 to 500 MHz with a total of 200 frequency steps. The loaded elements are shown in Fig. 9.2. The inductance value is $1\,\mu H$ (for each element). The resistance value is $200\,\Omega$ (for each element).

9.5.* For a 0.5 m long base-driven monopole on a 1 by 1 m finite ground plane (Fig. 9.5), estimate the effect of the resistive loading on the input impedance and the return loss. The resistance value is $100\,\Omega$. The load position is in the middle of the monopole. Use a frequency sweep of 150 to 1000 MHz with a total of 50 frequency steps. The antenna structure is generated using the script `monopole.m` from subdirectory `mesh`.

10

PATCH ANTENNAS

10.1. Introduction
10.2. Code Sequence
10.3. Model of the Probe Feed
10.4. Generation of the Antenna Structure
10.5. Input Impedance, Return Loss, and Radiation Pattern
10.6. Why Do We Need the Wide Patch?
10.7. A Practical Example
10.8. Dielectric Model
10.9. Accuracy of the Dielectric Model
10.10. Conclusions
       References
       Problems

10.1. INTRODUCTION

Patch or microstrip antennas [1,2] are probably the most widely used class of antennas today. They are important in many commercial applications, such as mobile radio and wireless communications. These antennas are low-profile and conformable to planar and nonplanar surfaces ranging from aircraft and rocket shapes to human bodies. The modern patch arrays perhaps constitute the most sophisticated and intriguing electromagnetic journey today. A comprehensive review of canonic patch/array shapes can be found in [2].

Compared to the other antenna shapes modeled in the preceding chapters, there are two new concepts that should be introduced when patch antennas

234 PATCH ANTENNAS

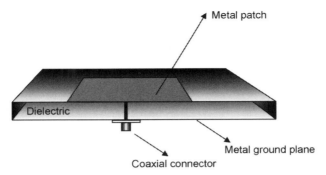

Figure 10.1. Typical probe feed for the microstrip antenna.

are studied. One concerns the supporting dielectric, and the other concerns the antenna feed. Patch antennas are very sensitive to the specific feeding method. Many of these methods exist [2].

Two groups of patch antennas may be considered: air-filled or low-dielectric ($\varepsilon_R \sim 1$) patch antennas and metal-dielectric antennas ($\varepsilon_R \sim 2\text{–}50$). The presence of a high-ε dielectric generally shifts the operation frequency to a considerably lower value and reduces its Q [1,3].

In this chapter we will study air-filled as well as low-ε finite-size patch antennas. The approximation of an electrically thin substrate [4] is used in the latter case. Such an approximation does not add much computational complexity to the MoM method used in the preceding chapters. Also the computational time remains nearly unchanged.

There are many configurations that can be used to feed patch antennas. The most popular are microstrip line feed, probe feed, and a proximity-coupled feed [1,2]. In this chapter we will mostly study the coaxial probe-feed patch antennas. This feed configuration is shown in Fig. 10.1. The way to model the probe feed is very similar to that for the base-driven monopole (Chapter 4).

The microstrip feed can be introduced in the algorithm as well. In this case the feeding edge should be the one of the strip edges. A printed dipole can be simulated in the same way. Then the feeding edge must the center edge of the dipole.

This chapter presents a simple patch antenna mesh generator that is used to create patch antenna structures in a matter of seconds. This generator employs the Matlab mouse function `ginput` to identify triangles belonging to the patch and the feed position. Multiple feeds and parasitic patches can be introduced.

10.2. CODE SEQUENCE

The code sequence is very similar to that of Chapter 7. The corresponding flowchart is shown in Fig. 10.2. To simplify the code, we omit 3D radiation

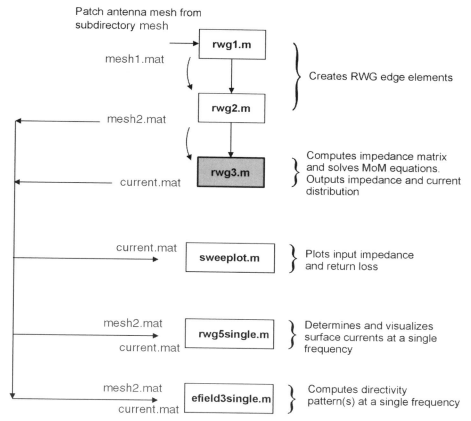

Figure 10.2. Flowchart of the code sequence of Chapter 10. The script containing the frequency loop is grayed.

patterns. Significant changes are made in the script rwg3.m where the code is modified to take the effect of dielectric into account. We will describe these changes in this chapter. The dielectric constant ε (electric permittivity of the dielectric substrate) should be specified at the beginning of this script.

The script rwg1.m introduces a new array EdgeIndicator that is used to identify edge elements belonging to the different parts of the patch antenna. The following values are available:

```
% 0 - metal ground plane
% 1 - feeding bottom edges
% 2 - connecting strip (probe feed)
% 3 - feeding top edges
% 4 - patch area
```

Indexes 1 and 3 corresponds to the edge elements that have one triangle, which belongs to the probe feed, and another triangle, which belongs to the ground plane or to the patch, respectively. Also all triangles of the structure now have the following identification numbers:

```
%t(4,:)=0 triangles of the metal ground plane
%t(4,:)=1 triangles of the probe feed
%t(4,:)=2 triangles of the patch
%t(4,:)=3 triangles of the upper boundary of dielectric
```

10.3. MODEL OF THE PROBE FEED

The thin-strip model of the probe feed is based on the algorithm of Chapter 4, and it essentially reproduces the model for the bottom-fed monopole. Figure 10.3 shows the probe feed model adapted to RWG edge elements. The starting point is a strip in Fig. 10.3a, which models the probe connector. In Fig. 10.3a, two RWG edge elements, having common triangle T, support the surface current $\mathbf{J}$ that is directed exactly along the strip axis. For a thin strip the radius of the equivalent cylindrical wire is given by Eq. (4.1), that is, $a_{eqv} = 0.25s$, where s is the strip width.

A model of the strip-plate junction is shown in Fig. 10.3b. There are two edge elements that have the common edge l_m. Such a model can connect two and more plates together as demonstrated in Fig. 10.3c.

For the base-driven probe, the feeding edge at junction l_m requires special treatment (Chapter 4). It is common to both the plate and the strip. Therefore two RWG elements correspond to the same edge. To separate these two elements it is convenient to "double" the junction edge, that is, just repeat it twice in the mesh code. The delta-function generator is then applied to edge l_m. The input impedance of the antenna is given by ratio of the common voltage through edge l_m (1 V) and the sum of the currents of two edge elements corresponding to l_m.

The feed identification is made in the script `rwg3.m`. This is done using the array `EdgeIndicator` introduced in Section 10.2. Since a bottom-driven

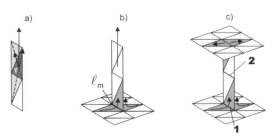

Figure 10.3. Model of the probe feed. (a) Single strip supporting vertical current; (b) strip-to-plate junction; (c) strip connecting two plates. 1, 2 indicate two possible positions of the feeding edge.

probe is mostly considered (EdgeIndicator==1), the following code is used to find the antenna impedance

```
Index=find(EdgeIndicator==1)
V(Index)=1*EdgeLength(Index);
. . .
GapCurrent(FF)=sum(I(Index).*EdgeLength(Index)');
GapVoltage(FF)=mean(V(Index)./EdgeLength(Index));
Impedance(FF)=GapVoltage(FF)/GapCurrent(FF)
```

The coaxial connector should be small enough compared to the antenna size. Otherwise, the model of the TEM magnetic frill should be employed [5] to properly model the coaxial line excitation.

10.4. GENERATION OF THE ANTENNA STRUCTURE

Before attempting the calculations for the patch antenna, we must first create the antenna structure. This is done in the scrip patchgenerator.m from the subdirectory mesh of the Matlab directory of this chapter. As an example we will execute that script and not change anything in the code. The Matlab plot is shown in Fig. 10.4a.

This figure shows the ground plane of a patch antenna. The other three figures follow the next three steps:

1. Use the Matlab mouse cross and the left mouse button to mark the white triangles shown in Fig. 10.4b. These triangles will belong to the patch(es).
2. Press return key after you are ready. The return key fixes the patch(es) structure. The picture will be updated as shown in Fig. 10.4c.
3. Use the Matlab mouse cross and the left mouse button to mark the white triangles shown in Fig. 10.4d. The edge between these triangles will belong to the feed(s). Press return key.

After these steps are completed, the antenna structure is created. It appears on the screen as shown in Fig. 10.5. The structure is the probe-fed finite patch antenna $10 \times 5 \times 1$ cm in size, with a rather narrow patch. This structure could be rotated to change the observation direction.

Looking through the script patchgenerator.m, one can see that it employs the Matlab function delaunay in order to create the ground plane mesh. Next it separates the patch triangles using the method of images in the ground plane and lifts them up to a certain height. After that, the probe feed is introduced as a common edge to two triangles identified at step 3. The script patchgenerator.m allows the following parameters to be changed

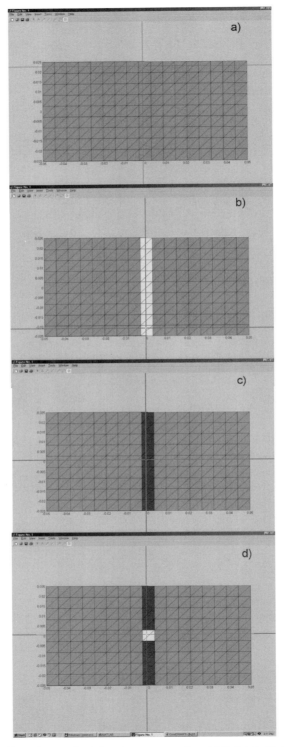

Figure 10.4. Four stages of patch antenna generation. Matlab mouse cross is seen.

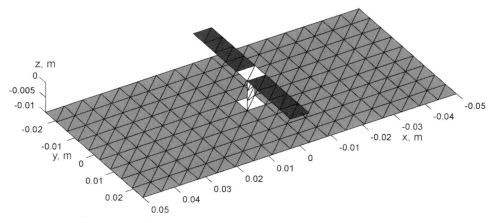

Figure 10.5. Antenna structure after execution `patchgenerator.m`.

```
%Separation distance between patch and ground plane
h=0.01;
L=0.10;    %Plate length (along the x-axis)
W=0.05;    %Plate width (along the y-axis)
Nx=17;     %Discretization parameter (length)
Ny=9;      %Discretization parameter (width)
%Number of rectangles of the feeding strip:
Number=3;
```

The script can introduce multiple patches and multiple feeds. The problems at the end of the chapter provide corresponding examples. The patch antenna's structure is not limited in size, but it is recommended that the total number of triangles does not exceed 4000.

Surprisingly the Matlab mouse's output has been found to be sensitive to the geometrical size of the structure. When the size of the structure is a small fraction of 1 (as it is usually the case), the mouse's output may stop working properly if the number of triangles exceeds 1000. Therefore it is necessary to rescale the antenna before the mesh generation.

Another problem with the script `patchgenerator.m` is that the width of the feeding strip is limited by the discretization accuracy of the ground plane. The two patch generators considered below are free of this limitation.

10.5. INPUT IMPEDANCE, RETURN LOSS, AND THE RADIATION PATTERN

Although the primitive patch antenna model introduced in the previous section (Fig. 10.5) is not very practical, it can readily be used to start the basic

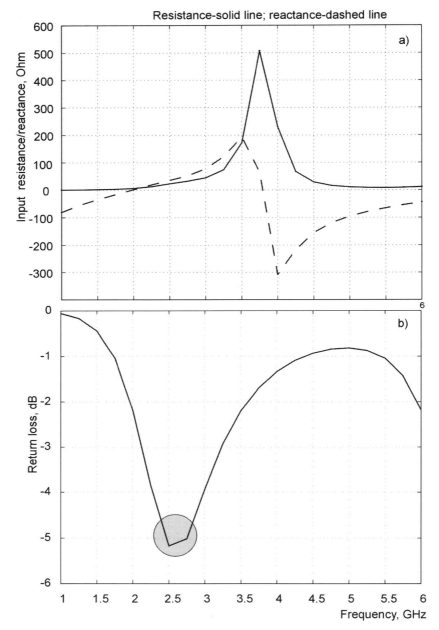

Figure 10.6. Input impedance (a) and return loss (b) of the simple patch antenna. The circle corresponds to the bandwidth area of the antenna.

calculations. First, the script `rwg1.m` would be run and reproduce Fig. 10.5. Next, `rwg2.m` would be executed. After the RWG edge elements are created, the script `rwg3.m` would be run. By default this script does a frequency loop of 21 points over the band from 1 to 6 GHz. The default dielectric constant (relative dielectric permittivity) of the substrate is 1. The execution time for the script `rwg3.m` should be about 6 minutes on a Pentium IV processor with Intel™ motherboard.

After the script `rwg3.m` is executed, the script `sweeplot.m` provides us with the input impedance and return loss of the antenna over the desired frequency band. Figure 10.6 shows the input impedance of the antenna and its return loss. The solid line denotes the resistance (impedance real part), and the dashed line corresponds to the reactance (impedance imaginary part).

The input impedance exhibits typical "resonance" behavior with the resistance peak [1, p. 763]. Such behavior is very common for other resonant antennas, as we learned in Chapter 7, and not only for the patch antenna. The return loss has a minimum of approximately −5 dB at 2.5 GHz. When a finer frequency step is employed, the minimum value drops to approximately −8 dB. To ensure proper load matching, the antenna center frequency must therefore be close to 2.5 GHz.

The radiation patterns are obtained using the script `efield3single.m`. They are shown in Fig. 10.7. It is seen that the present patch antenna performs rather poorly. In particular, the zero radiated field occurs on the antenna's axis. This is a weak design since the microstrip patch should have its maximum radiation normal to the patch (broadside radiator) [1].

There are two reasons why the broadside radiation is missing. First, the structure is symmetric. Since the currents on the "dipole patch" are oppositely directed, their contributions cancel each other in the far field. Figure 10.8 shows the surface current distribution on the patch obtained using the script `rwg5single.m` at 2.5 GHz.

Second, the patch itself is too narrow to radiate effectively. In conventional patch antennas it is not the patch but the regions along the patch's edges that radiate the electromagnetic signal [1]. In these regions the E-field lines are fringed. The region beneath the patch is basically a high-Q cavity resonator with an approximately uniform electric field. So, in order to achieve better radiation, we have to make the patch wider and enlarge its radiating circumference.

10.6. WHY DO WE NEED A WIDE PATCH?

In this section we modify the preceding example in order to make the patch antenna work properly. The new antenna structure is shown in Fig. 10.9. The structure is again obtained running the script `patchgenerator.m`. We do not have to change anything in the code. However, in contrast to the previous example, we increase the patch size (the patch now contains 9×5 rectangles).

242 PATCH ANTENNAS

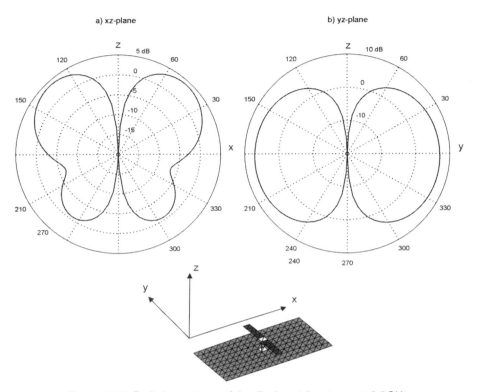

Figure 10.7. Radiation patterns of the dipole patch antenna at 2.5 GHz.

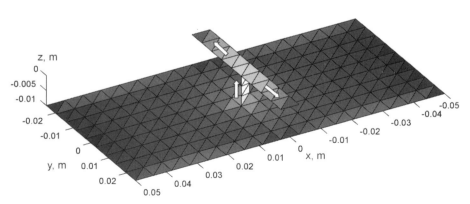

Figure 10.8. Surface current distribution of the dipole patch antenna at 2.5 GHz. The white color corresponds to larger magnitudes.

WHY DO WE NEED A WIDE PATCH? 243

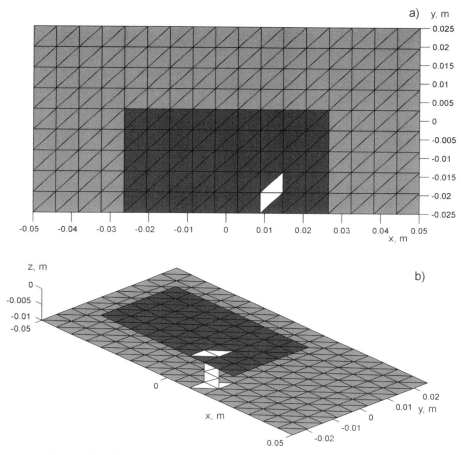

Figure 10.9. Patch antenna structure after execution patchgenerator.m.

Also the antenna feed is shifted toward one of the patch edges. Both of these changes are critical for the antenna's performance. The interested reader should follow exactly the steps of Section 10.4 to create the patch antenna shown in Fig. 10.9.

The main code sequence rwg1.m, rwg2.m, and rwg3.m is executed. The result for the return loss is shown in Fig. 10.10a. We see that the minimum of return loss again occurs at 2.5 GHz. This is the center frequency of the antenna. The load matching at 2.5 GHz is excellent, so the return loss drops to −16 dB.

Two radiation patterns of the antenna are shown in Fig 10.10b. Clearly, the present patch antenna performs very well as a broadside radiator with approximately a 60° beamwidth.

We know from the literature [1,2] that real input impedances are obtained when the patch dimensions are roughly $\lambda/2$ on a side. In Fig. 10.9, the patch

244 PATCH ANTENNAS

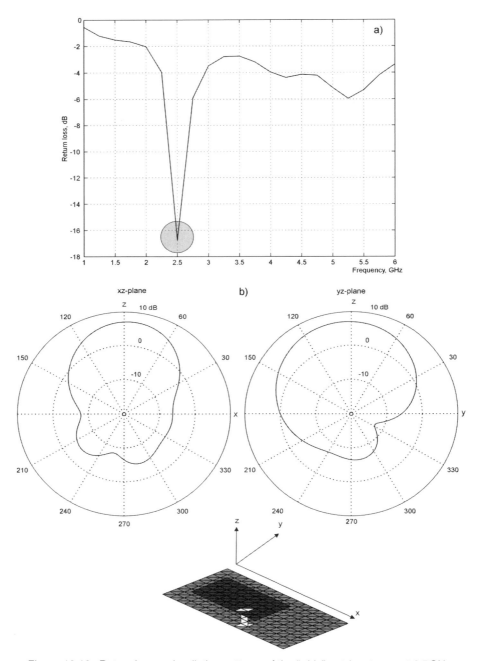

Figure 10.10. Return loss and radiation patterns of the "wide" patch antenna at 2.5 GHz.

length is approximately 5.6 cm, which corresponds to $\lambda = 11.2$ cm and a center frequency of $f = 2.7$ GHz. This value is in a good agreement with the frequency $f = 2.5$ GHz which corresponds to the minimum of the return loss in Fig. 10.10a.

10.7. A PRACTICAL EXAMPLE

Now we consider a practical example of the dual-band *E*-shaped patch antenna studied in [7]. The air-filled patch antenna operates at 1.9 as well as 2.4 GHz [7, sec. III]. The patch height is 0.015 m. The antenna is fed by a coaxial probe. The probe location corresponds to the center of the common edge of two small white triangles in Fig. 10.11a. A large but finite ground plane is assumed according to Ref. [7, fig. 1].

The antenna has a large ground plane and a slotted patch (Fig. 10.11). Therefore the mesh of equal triangles created by patchgenerator.m will be too large. It is not appropriate for our purposes. We need a nonuniform mesh that has the finest resolution close to the patch slots as well as to the feed.

The initial 2D antenna template is created using the mesh generator of the Matlab PDE toolbox as shown in Fig. 10.11a. The corresponding PDE file is eshape.m in subdirectory mesh. If you have the PDE Toolbox installed on your machine, you can run this file using the Matlab command prompt to see the template. The large rectangle created by rectangle tool corresponds to the ground plane. An E-shaped polygon inside the rectangle is created using polygon tool, and it corresponds to the patch. A small rectangle within the patch is also created to refine the mesh close to the feed.

After we export the mesh to the main Matlab workspace (use export mesh and then press ok button) and save the result as eshape.mat, the array of triangles will have in its last row the following display:

```
t(4,:)   =1            triangles of the metal ground plane
t(4,:)   =2 (or 3)     triangles of the patch
```

The Matlab script pachgenerator1.m in subdirectory mesh is intended to work with the PDE toolbox. It reads the binary file eshape.mat and creates the 3D patch antenna based on the existing triangle indication. It is not necessary to identify the patch shape using mouse. However, the mouse input still exists to identify the probe feed by marking two white triangles somewhere on the patch. The output of the script is shown in Fig. 10.11b. The width of the feeding strip is 0.0015 m. The structure has totally 1395 edge elements.

Scripts patchgenerator.m and patchgenerator1.m are two alternatives to the patch antenna mesh generation. Whereas the first script is rather a "quick and dirty" mesh generator, the second allows more precise adjustments in the equivalent size of the probe feed.

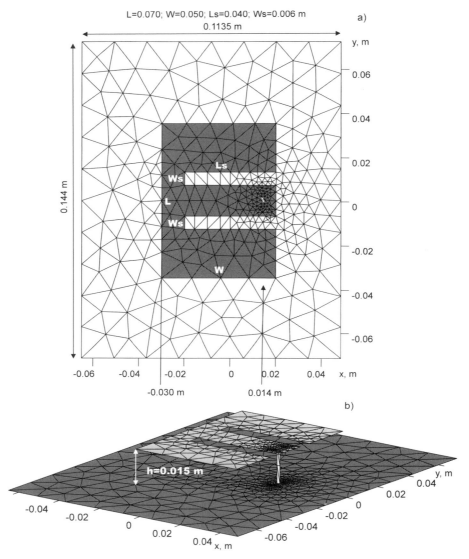

Figure 10.11. Patch antenna from Ref. [7] to scale. (a) Output of the PDE toolbox; (b) output of `patchgenerator1`.

Next, the main code sequence `rwg1.m`, `rwg2.m`, and `rwg3.m` is executed. Calculations are done over the band 1.5 to 3.0 GHz. Two models are considered: (1) feeding edge at the bottom edge of the strip (bottom feed 1 in Fig. 10.3c); and (2) feeding edge at the middle edge of the strip (center feed 2 in Fig. 10.3c). The results for the return loss are presented in Fig. 10.12. The experimental data [7] (dashed curve) and the HP-HFSS simulation (stars) performed in [7] are also given.

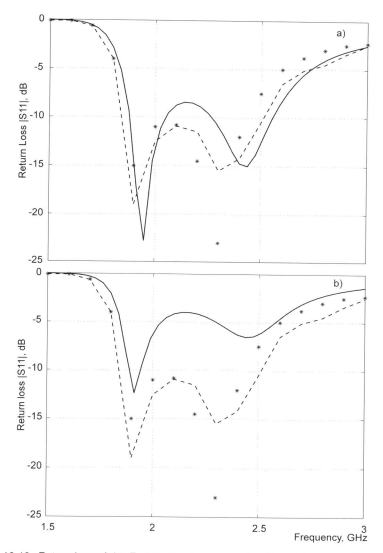

Figure 10.12. Return loss of the E-shaped antenna as a function of frequency. Dashed line: experiment [7]; stars: HFSS simulation [7]; solid line: present simulation. (a) Bottom feeding edge; (b) center feeding edge.

In Fig. 10.12a we see that the bottom feed provides good agreement between the present results and experiments/simulations done in [7]. At the same time the center feed model (Fig. 10.12b) fails in almost the entire frequency domain. Very similar results were obtained for other orientations and sizes of the feeding edge on the patch plane. Thus the bottom feeding edge with two adjacent edge elements reasonably well describes the probe excitation of the patch antenna. The coaxial line diameter should be small enough compared to the antenna size. Otherwise, the model of the TEM magnetic frill should be employed [5] for coaxial line excitation.

10.8. DIELECTRIC MODEL

The full-wave analysis of patch antennas via MoM involves substantial computational complexity. For infinite substrates, the Green's function of the grounded dielectric slab has to be evaluated carefully [8–10]. For finite ground planes, different sets of basis functions are necessary either for the volume/surface formulation or for the electric and magnetic currents of the EFIE [11,12].

The simple moment method solution presented here must be viewed as an approximate solution valid for electrically thin dielectric slabs. It originates from the paper of Newman and Tulyathan [4]. The finite dielectric slab is removed and replaced by the equivalent volume polarization currents **J** (A/m²),

$$\mathbf{J} = j\omega\varepsilon(\varepsilon_R - 1)\mathbf{E} \qquad (10.1)$$

where ε is dielectric permittivity of vacuum, ε_R is relative permittivity, and **E** is the actual electric field in the slab. The dielectric volume is divided into triangular volume elements as shown in Fig. 10.13. Triangles of the surface mesh are used for that purpose. The electric field within each volume element is assumed to have only one significant vertical component. This component is defined by the surface charges at the top patch and the bottom patch of every volume element

$$\mathbf{E} = \begin{cases} \hat{\mathbf{z}} \dfrac{1}{2\varepsilon}(\rho_s^+ - \rho_s^-) & \text{between ground plane and patch (Fig. 10.13}a) \\ \hat{\mathbf{z}} \dfrac{1}{2\varepsilon}(\rho_s^+ - 0) & \text{between ground plane and vacuum (Fig. 10.13}b) \end{cases} \qquad (10.2)$$

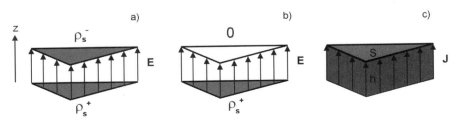

Figure 10.13. Quasi-static electric field distribution in a dielectric substrate.

where $\rho_S^{\pm}$ is the surface charge density on the metal ground plane and the metal patch. Equation (10.2) is the well-known quasi-static relation for the parallel-plate capacitor. The surface charges are those for the RWG elements on the metal surface.

The volume polarization currents (10.1) alter the electric field on the metal surface. They also interact with each other through the radiated electric field. The radiated electric field of a single volume element shown in Fig. 10.13c is obtained using the dipole model (see Chapter 3). The dipole model replaces the current in the volume element by a finite-length vertical dipole of height h. For better accuracy, several dipoles can be considered that are uniformly distributed over the cross section S of the volume element.

The modified moment equations take the form (index S refers to metal and index D to dielectric)

$$\begin{bmatrix} Z_{SS} & -Z_{DS} \\ Z_{SD} & Z_{DD} - I \end{bmatrix} \begin{bmatrix} \mathbf{I}_S \\ \mathbf{I}_D \end{bmatrix} = \begin{bmatrix} \mathbf{V} \\ 0 \end{bmatrix} \quad (10.3)$$

where $\mathbf{I}_D$ is the vector of unknown volume currents. Matrix Z_{DS} is filled using the dipole model. Its mn-element is proportional to the electric field on RWG element m due to volume element n. Function point_.m is used to calculate the electric field of a finite-length dipole. Matrix Z_{SD} is filled using the capacitor model (10.2). Its mn-element is proportional to the electric field in volume element m due to RWG element n. Square matrix Z_{DD} (self-interaction matrix) is filled using the dipole model as well.

These changes are incorporated into the script rwg3.m where the matrixes Z_{SS}, Z_{DS}, Z_{SD}, and Z_{DD} are calculated separately. The total execution time is approximately 100% higher than the execution time of the pure metallic model. The execution time can be reduced using the code vectorization.

To pursue the full-wave solution, the lateral electric current in the dielectric must be supported. This can be done by doubling the ground layer of RWG edge elements and lifting it up by $h/2$, where h is the patch height. Such an intermediate layer is capable of supporting any lateral current distribution in the dielectric. The full-wave dielectric solution for the patch antennas using RWG edge elements will be announced on the book website.

10.9. ACCURACY OF THE DIELECTRIC MODEL

Figure 10.14 shows a dielectric patch antenna setup to scale. This setup is a modified project DEMO-366 of WIPL-D software [5]. The modifications imply that (1) patch thickness is reduced to minimum, (2) the feeding point is moved from the patch boundary to the inner point (0.1 0.166) of the patch, and (3) the patch height is increased to 0.05 m. The probe radius 0.004 m is chosen, and it corresponds to a strip width of 0.016 m. WIPL-D presumably uses "theorem of transfer of excitation" [5, p. 58] and the center-fed wire segment to model

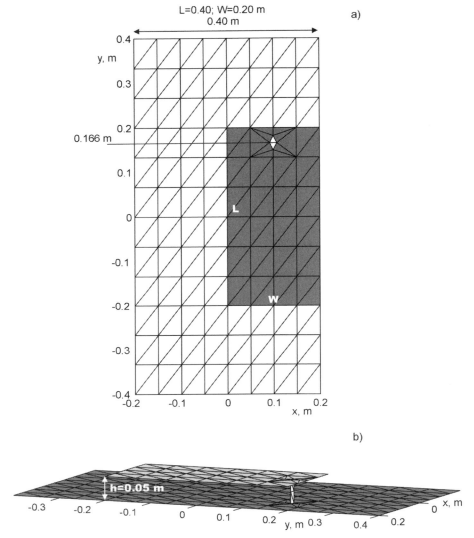

Figure 10.14. Dielectric-filled patch antenna setup to scale. (a) Intermediate result after identifying the feed position; (b) output of patchgenerator2.

the probe feed of a patch antenna. Therefore, for the purposes of comparison, the center feed model of Fig 10.3c is used here. It is programmed in the script rwg3.m just below the code for the base-driven probe.

The initial uniform 2D mesh is created with the help of the script patchgenerator2.m in the subdirectory mesh. This script essentially repeats the script patchgenerator.m. However, it manually introduces additional vertexes in order to refine the uniform mesh X, Y in the vicinity of the probe feed and control the probe thickness. This is done by adding four points to the existing vertexes. The addition to the code of `patchgenerator.m` may have the following form:

```
%Identify probe feed edge
x=[0.092  0.108];
y=[0.166  0.166];
X=[X x];
Y=[Y,y];
x1=[mean(x)  mean(x)];
y1=mean(y)+2*[max(x)-mean(x)  min(x)-mean(x)];
X=[X x1];
Y=[Y,y1];
```

The rest of the script remains unchanged. The result is seen in Fig. 10.14b. The total number of RWG edge elements (359) matches the number of unknowns in the corresponding WIPL-D model (variable from 310 to 474).

Antenna calculations are done over the band 200 to 400 MHz. The results for the return loss are presented in Fig. 10.15a to c for three different values of the dielectric constant ε_R = 1.0, 1.5, 2.6. The solid line corresponds to the present code, whereas stars denote the WIPL-D results.

Reasonably good agreement is observed in Fig. 10.15. We can track the evolution and decay of the fundamental antenna resonance shown by black arrows as ε_R increases. Also noticeable is a significant disagreement for the pure metallic model in Fig. 10.15a. We believe that this disagreement is due to insufficient discretization accuracy close to the antenna feed.

10.10. CONCLUSIONS

We have presented the code sequence for the air-filled patch antennas and for the patch antennas in the approximation of the electrically thin substrate. These codes were tested for a few examples.

The patch antenna structure can be created using one of the scripts `patchgenerator/patchgenerator1/patchgenerator2.m` in the subdirectory `mesh`. These scripts are described in the text. To enable faster structure processing, all generator scripts include mouse output. Multiple feeds and a patch antenna array can be created using these scripts.

The total execution time of the model is approximately 100% higher than the execution time of the pure metallic model. To pursue the full-wave solution, the lateral electric current in the dielectric must be supported. This can be done by doubling the ground layer of RWG edge elements and lifting it up by $h/2$, where h is the patch height. Such an intermediate layer is capable of supporting any lateral current distribution in the dielectric. No other surface or volume discretization becomes necessary. The interested reader can modify

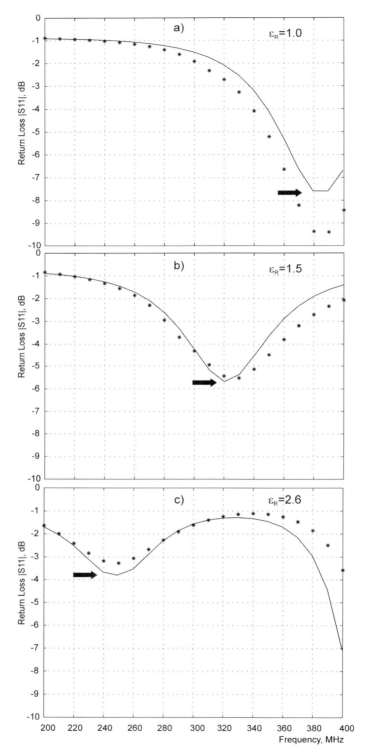

Figure 10.15. Return loss as a function of frequency for the tested patch antenna. Stars: WIPL-D simulation [5]; solid line: present model. The arrows show the evolution of the fundamental resonance as the dielectric constant increases.

the source code accordingly. The full-wave dielectric solution for the patch antennas using RWG edge elements will be announced on the book's website.

REFERENCES

1. C. A. Balanis. *Antenna Theory: Analysis and Design*, 2nd ed., ch. 14. Wiley, New York, 1997.
2. J. R. James and P. S. Hall. *Handbook of Microstrip Antennas*, Vols. 1, 2. Peter Peregrinus Ltd., London, 1989.
3. J.-F. Zücher and F. E. Gardiol. *Broadband Patch Antennas*, Artech House, Norwood, MA, 1995.
4. E. H. Newman and P. Tulyathan. Analysis of microstrip antennas using moment method. *IEEE Trans. Antennas and Propagation*, 29 (1): 47–53, 1981.
5. B. M. Kolundžija, J. S. Ognjanović, and T. K. Sarkar. *WIPL-D: Electromagnetic Modeling of Composite Metallic and Dielectric Structures*. Artech House, Norwood, MA, 2000.
6. D. M. Pozar. *Microwave Engineering*, 2nd ed. Wiley, New York, 1998, p. 67.
7. F. Yang, X.-X. Zhang, X. Ye, and Y. Rahmat-Samii. Wide-band E-shaped patch antennas for wireless communications. *IEEE Trans. Antennas and Propagation*, 49 (7): 1094–1100, 2001.
8. M. Marin, S. Barkeshli, and P. H. Pathak. Efficient analysis of planar mictrostrip geometries using a closed-form asymptotic representation of the grounded dielectric slab Green's function. *IEEE Trans. Microwave Theory and Techniques*, 37 (4): 669–679, 1989.
9. R. Kipp and C. H. Chan. Triangular-domain basis functions for full-wave analysis of microstrip discontinuities. *IEEE Trans. Microwave Theory and Techniques*, 41 (6/7): 1187–1194, 1993.
10. L. Tarricone, M. Mongiardo, and F. Cervelli. A quasi-onedimensional integration technique for the analysis of planar microstrip circuits via MPIE/MoM. *IEEE Trans. Microwave Theory and Techniques*, 49 (3): 517–523, 2001.
11. T. K. Sarkar, S. M. Rao, and A. R. Djordjević. Electromagnetic scattering and radiation from finite microstrip structures. *IEEE Trans. Microwave Theory and Techniques*, 38 (11): 1568–1575, 1990.
12. B. G. Salman and A. McCowen. The CFIE technique applied to finite-size planar and non-planar microstrip antenna. *Computation in Electromagnetics* (Conf. Publ. No. 420), pp. 338–341, 1996.
13. D. Sievenpiper, L. Zhang, R. F. Jimenez Broas, N. G. Alexopolous, and Eli Yablonovitch. High impedance electromagnetic surfaces with a forbidden frequency band. *IEEE Trans. Microwave Theory and Techniques*, 47 (11): 2059–2074, 1999.
14. D. Sievenpiper, H. P. Hsu, J. Schaffner, G. Tangonan, R. Garcia, and S. Ontiveros. Low-profile, four sector diversity antenna on high-impedance ground plane. *Electronics Letters*, 36 (16): 1343–1345, 2000.
15. R. F. Jimenez Broas, D. Sievenpiper, and Eli Yablonovitch. A high-impedance ground plane applied to a cellphone handset geometry. *IEEE Trans. Microwave Theory and Techniques*, 49 (7) 1262–1265, 2001.

254 PATCH ANTENNAS

PROBLEMS

10.1. Using the script `patchgenerator.m`:
  a. Create the antenna structure shown in Fig. 10.16a.
  b. Create the antenna structure shown in Fig. 10.16b.
  c. Create the antenna structure shown in Fig. 10.16c.
  d. Create the antenna structure shown in Fig. 10.16d.

10.2. Using the script `patchgenerator2.m`:
  a. Create the patch array shown in Fig. 10.17a.
  b. Create the patch array shown in Fig. 10.17b.

10.3. For the antenna shown in Fig. 10.16a calculate the return loss over the band 0.5 to 5 GHz. Use 41 frequency points per band. Based on those calculations:
  a. Find the center frequency of the antenna.
  b. Plot radiation patterns and the surface current distribution at the center frequency.
  c. Is the patch antenna broadside or end-fire?

10.4. Repeat Problem 10.3 when the patch height is two times smaller.

10.5. For the antenna shown in Fig. 10.16b calculate the return loss over the band 0.5 to 6 GHz. Use 41 frequency points per band. Based on those calculations:
  a. Find the center frequency of the antenna.
  b. Plot radiation patterns and the surface current distribution at the center frequency.

10.6. For the antenna shown in Fig. 10.16c calculate the return loss over the band 0.5 to 6 GHz. Use 41 frequency points per band. Based on those calculations:
  a. Find the center frequency of the antenna.
  b. Estimate the antenna bandwidth.
  c. Plot radiation patterns and the surface current distribution at the center frequency.
  Is the patch antenna broadside or end-fire?

10.7. The antenna in Fig. 10.16d is a series-connected patch array [1, p. 773]. Find the center frequency of the antenna over the band 1 to 6 GHz. Use 81 frequency points per band. Plot the radiation patterns at that frequency.

10.8. Use the script `patchgenerator2.m` to create the antenna structure shown in Fig. 10.14. Change the antenna height to 0.03 m, and calculate and plot the return loss over the band 200 to 400 GHz at

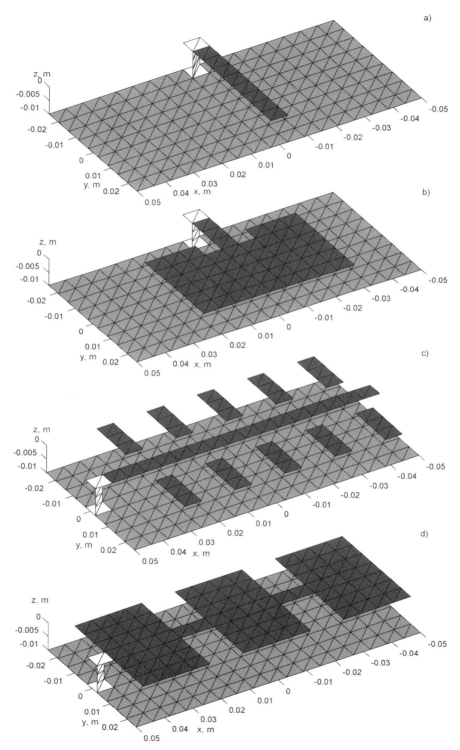

Figure 10.16. Various patch antenna configurations.

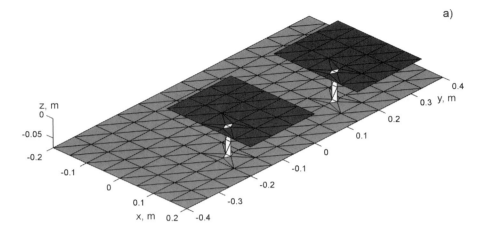

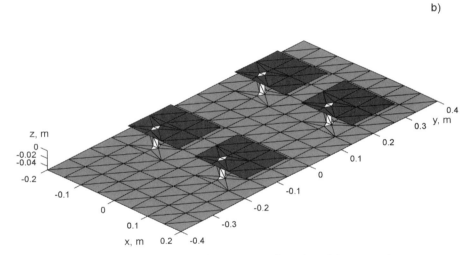

Figure 10.17. Various patch antenna configurations (planar array).

a. $\varepsilon_R = 1.0$
b. $\varepsilon_R = 1.5$
c. $\varepsilon_R = 2.0$

Use 41 frequency points per band.

10.9. A high-impedance surface has recently been proposed that improves the characteristics of patch antennas [13–15]. The surface has a typical "mushroom" structure as shown in Fig. 10.18. Create and plot the high-impedance surface shown in Fig. 10.18.

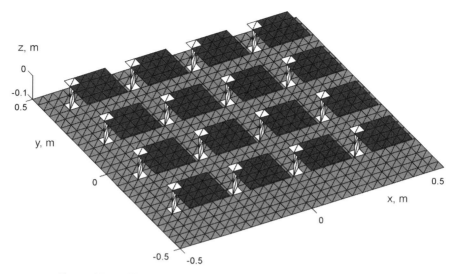

Figure 10.18. High-impedance "mushroom" surface similar to [13–15].

10.10.* The patch antenna array shown in Fig. 10.18 is created using `patch-generator.m`. It has multiple feeds (one feed per patch). All feeds have `EdgeIndicator=1` (see Section 10.2) and cannot be separated in the present code. Change the code of the script `rwg3.m` in order to make only one certain feed active. Attach the corresponding code modification and test it. Hint: Use a simplified model with only two patches to test the code.

APPENDIX A

OTHER TRIANGULAR MESHES

A.1. Plate Mesh
A.2. Cube Mesh and Volume Metal Grid
A.3. Cavity Model
A.4. Strip Mesh
A.5. Spherical and Cylindrical Structures

A.1. PLATE MESH

The mesh for a plate of arbitrary length/width having an arbitrary discretization accuracy is created running the script plate.m. This is essentially equivalent to the script strip.m used in the main text. The script outputs the mesh into binary file plate.mat. Run this script to see the result.

A.2. CUBE MESH AND VOLUME METAL GRID

The code volumemesh.m is intended to create rectangular volume grid meshes having arbitrary discretization accuracy. It creates a mesh cube as a degenerate case. Although it is being extensively tested, this code has not yet been properly optimized. Therefore it is very lengthy and slow. The script uses function strip_.m, which is *not* equivalent to the code strip.m of the main text. The input parameters of the script are illustrated in Fig. A.1.

260   OTHER TRIANGULAR MESHES

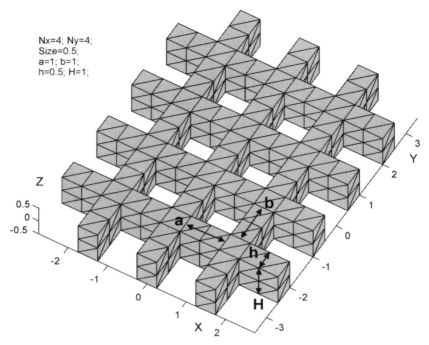

Figure A.1. Input parameters for the script `volumemesh.m` (left upper corner) and the resulting mesh.

Figure A.2 shows the script output for two other sets of input parameters. These include a cube (Nx=Ny=1) and a 9×10 thick metal grid (Nx=Ny=10). Strictly speaking, the number of bars perpendicular to the x-axis is not equal to Nx but to Nx-1. Variable `size` controls the mesh grid size.

A.3. CAVITY MODEL

The code `volumemesh1.m` is a modification of the code `volumemesh.m`. It removes the upper layer of triangles and creates an "open" volume metal mesh. The degenerate case of the cube thus corresponds to an open rectangular cavity. Figure A.3 shows two possible cavity meshes.

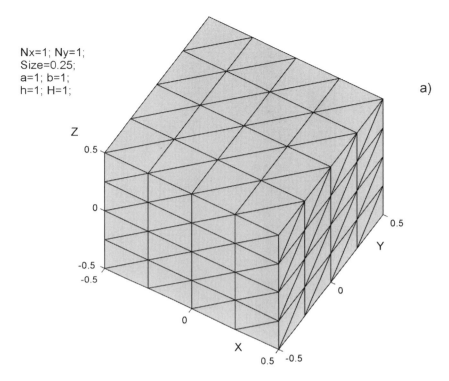

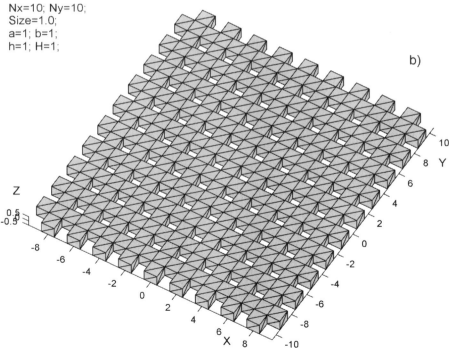

Figure A.2. Input parameters for the script `volumemesh.m` (left upper corner) and the resulting mesh. (a) Metal cube; (b) metal grid.

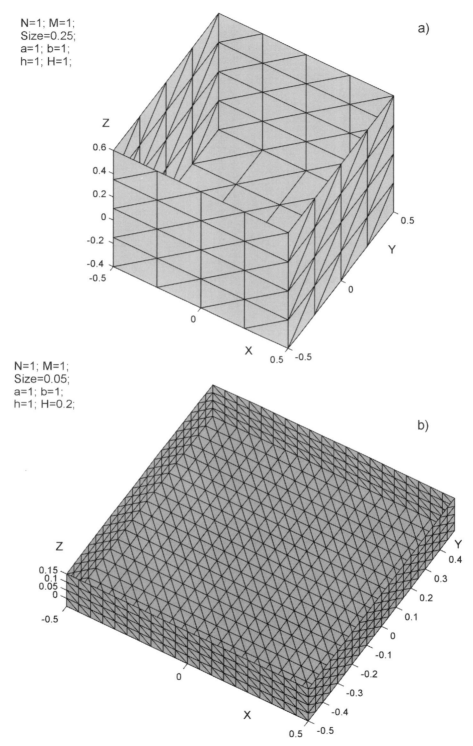

Figure A.3. Input parameters for the script `volumemesh1.m` (left upper corner) and the resulting mesh. (a) Deep rectangular cavity; (b) shallow cavity.

A.4. STRIP MESH

The code `volumemesh1.m` at $H = 0$ outputs planar meshes of intercepting strips with various cell sizes and strip widths. Figure A.4 provides an example. Note that these planar meshes can be combined together to form a volumetric lattice of strip layers.

A.5. SPHERIC AND CYLINDRICAL STRUCTURES

Sphericae and cylindrical structures (`sphere.mat`, `sphere1.mat`, `cylinder.mat`) are imported to Matlab in ASCII format. They are shown in Fig. A.5. There are no special mesh generators for these structures. Structure `sphere.mat` has the radius of 1 m and 500 triangles. Structure `sphere1.mat` has the radius of 1 m and 8000 triangles. The finite 10:1 cylinder `cylinder.mat` has 340 triangles.

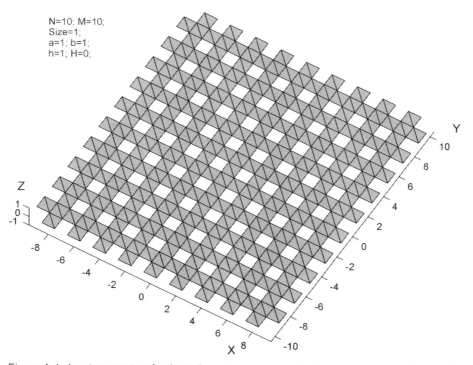

Figure A.4. Input parameters for the script `volumemesh1.m` (left upper corner) and the resulting planar strip mesh.

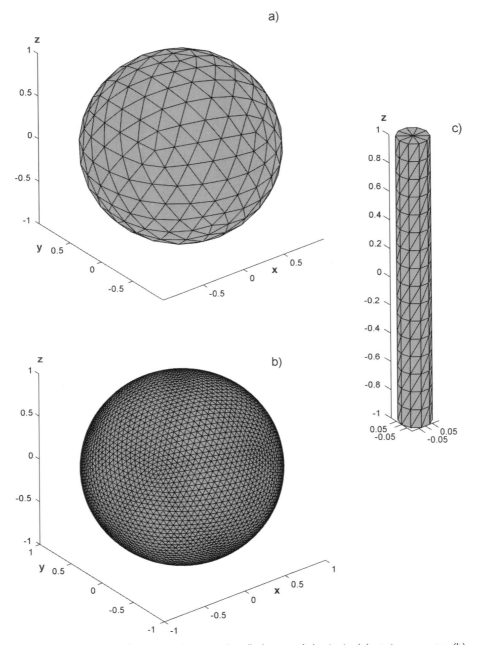

Figure A.5. Meshes for the sphere and cylinder used in text. (a) Sphere.mat; (b) sphere1.mat; (c) cylinder.mat.

APPENDIX B

IMPEDANCE MATRIX CALCULATION

B.1. Impedance Matrix Calculations
B.2. Self-coupling Terms
   References

B.1. IMPEDANCE MATRIX CALCULATIONS

Below we present the algorithm for the impedance matrix calculation employed in the present text. The algorithm essentially utilizes "the face-pair" method discussed in [1] but is not equivalent to it. It is based on the summation of contributions due to integration over separate triangles. The names of variables are kept as close as possible to the original variables used in the source code for the impedance matrix (function `impmet.m`).

A triangle T_p, $p = [1:$ `TrianglesTotal`$]$ may be a part of a few edge elements that include T_p as the "plus" triangle. The numbers of those elements are stored in the array `Plus`:

$$\text{Plus} \quad =\text{find(TrianglePlus-p==0)}; \tag{B1}$$

Note that the array `Plus` may have any length from 0 to 3. The same triangle T_p, $p = [1:$ `TrianglesTotal`$]$ may be a part of a few other edge elements that include T_p as the "minus" triangle. The numbers of those elements are stored in the array `Minus`:

$$\text{Minus} \quad =\text{find(TriangleMinus-p==0)}; \tag{B2}$$

The array `Minus` may have any length from 0 to 3.

266 IMPEDANCE MATRIX CALCULATION

It can be seen from Eqs. (2.2) and (2.4) of Chapter 2 that only the following elements of the impedance matrix:

$$Z_{[1:\text{EdgesTotal}],\text{Plus}} \tag{B3}$$

and

$$Z_{[1:\text{EdgesTotal}],\text{Minus}} \tag{B4}$$

include the integrals over triangle T_p. Indeed, along with the integrals over T_p, they may include integrals over other triangles.

It is therefore suggested to find all possible integrals over T_p first and then accumulate these in the appropriate elements of matrix Z given by (B3) and (B4). Hence, the "contribution" of triangle T_p into the impedance matrix will be completed.

Next, triangle T_{p+1} is considered. Contributions associated with the integrals over T_{p+1} are accumulated in the appropriate elements of Z in the same way. Further, the process continues as a loop over all triangles of the structure. Once integrals over all triangles are calculated, the impedance matrix is filled.

In particular, the integrals over T_p are accumulated in the following way (combination of Eqs. (2.2) and (2.4) of Chapter 2):

$$Z_{m,n=\text{Plus}(k)} = Z_{m,n=\text{Plus}(k)}$$

$$+ \left(\frac{j\omega l_m}{4}\right)\left(\frac{\mu}{4\pi}\right)\frac{l_n}{A_n^+} \int_{T_n^+} [g_m^+ \boldsymbol{\rho}_m^{c+} \cdot \boldsymbol{\rho}_n^+(\mathbf{r}') + g_m^- \boldsymbol{\rho}_m^{c-} \cdot \boldsymbol{\rho}_n^+(\mathbf{r}')] \cdot dS'$$

$$+ \left(\frac{l_m}{4\pi j\omega\varepsilon}\right)\frac{l_n}{A_n^+} \int_{T_n^+} [g_m^+ - g_m^-] dS'$$

$$Z_{m,n=\text{Minus}(k)} = Z_{m,n=\text{Minus}(k)}$$

$$+ \left(\frac{j\omega l_m}{4}\right)\left(\frac{\mu}{4\pi}\right)\frac{l_n}{A_n^-} \int_{T_n^-} [g_m^+ \boldsymbol{\rho}_m^{c+} \cdot \boldsymbol{\rho}_n^-(\mathbf{r}') + g_m^- \boldsymbol{\rho}_m^{c-} \cdot \boldsymbol{\rho}_n^-(\mathbf{r}')] \cdot dS'$$

$$- \left(\frac{l_m}{4\pi j\omega\varepsilon}\right)\frac{l_n}{A_n^-} \int_{T_n^-} [g_m^+ - g_m^-] dS'$$

$$g_m^\pm = \frac{e^{-jk|\mathbf{r}_m^{c\pm} - \mathbf{r}'|}}{|\mathbf{r}_m^{c\pm} - \mathbf{r}'|} \tag{B5}$$

Eqs. (B5) are programmed in the function impmet.m. Functions $g, \boldsymbol{\rho}, \boldsymbol{\rho}^{c\pm}$ are given by Matlab arrays g, RHO__Plus/Minus, and PHO_P/M. The latter (contemporary) arrays are replicated versions of the arrays RHO_Plus/Minus over nine triangle sub-points.

B.2. SELF-COUPLING TERMS

One must realize that the 9-point numerical quadrature (2.1) of Chapter 2 is a simple but not necessarily the most accurate method of the impedance matrix calculation, especially for the diagonal terms. More accurate numerical quadrature or an analytical formula may be employed.

The accuracy of quadrature (2.1) for the diagonal terms of the impedance matrix is discussed below. Here, all integrals containing singularity $1/r$ within the triangle (the so-called self-coupling terms) are approximated using the analytical formula given in [2]. All other non-singular integrals still use the 9-point quadrature. Surprisingly, the mixed approach gives the results that are very close to the results of Chapter 2. The reason for this is likely the integrable character of the singularity $1/|\mathbf{r}|$ on the plane. Interested readers may use the algorithm of Appendix B instead of the algorithm of Chapter 2. The scripts rwg2.m, rwg3.m, and impmet.m should be replaced by those from the Matlab directory of Appendix B.

The impedance matrix introduced in [1] originally included the following double integrals:

$$\int_{T_p} a(\mathbf{r}) \int_{T_q} b(\mathbf{r}') \frac{e^{-jk|\mathbf{r}-\mathbf{r}'|}}{|\mathbf{r}-\mathbf{r}'|} dS dS' \tag{B6}$$

where a and b may be either $\boldsymbol{\rho}$ or a constant. The next step of Ref. 1 was to eliminate integration over $\mathbf{r}$ by introducing the one-point approximation, at the triangle centroid $\mathbf{r}^c$. This is the way in which the impedance matrix in Chapter 2 (and Eq. (B5)) was obtained. In order to calculate the self-coupling terms (when $p = q$ in Eq. (B6)) we will start with the original expression Eq. (B6). When the triangle size is small compared to wavelength, one can use Taylor series expansion

$$e^{-jk|\mathbf{r}-\mathbf{r}'|} \approx 1 - jk|\mathbf{r}-\mathbf{r}'| \tag{B7}$$

Therefore

$$\int_{T_p} a(\mathbf{r}) \int_{T_p} b(\mathbf{r}') \frac{e^{-jk|\mathbf{r}-\mathbf{r}'|}}{|\mathbf{r}-\mathbf{r}'|} dS dS' \approx -jk \int_{T_p} a(\mathbf{r}) \int_{T_p} b(\mathbf{r}') dS dS' + \int_{T_p} a(\mathbf{r}) \int_{T_p} b(\mathbf{r}') \frac{1}{|\mathbf{r}-\mathbf{r}'|} dS dS' \tag{B8}$$

Further, the function $a(\mathbf{r})$ is replaced by its value at the triangle centroid $\mathbf{r}^c$. The function $b(\mathbf{r}')$ may be replaced either by its value at the triangle centroid or by its averaged value over the triangle area. Thus, one has to calculate the potential integral in the last term on the right-hand side of Eq. (B8). According to [2]

$$\frac{1}{4A_p^2}\int_{T_p}\int_{T_p}\frac{1}{|\mathbf{r}-\mathbf{r'}|}dSdS' =$$

$$\frac{1}{6\sqrt{a}}\ln\left[\frac{(a-b+\sqrt{ad})(b+\sqrt{ac})}{(-a+b+\sqrt{ad})(-b+\sqrt{ac})}\right]+$$

$$\frac{1}{6\sqrt{c}}\ln\left[\frac{(-b+c+\sqrt{cd})(b+\sqrt{ac})}{(b-c+\sqrt{cd})(-b+\sqrt{ac})}\right]+\frac{1}{6\sqrt{d}}\ln\left[\frac{(a-b+\sqrt{ad})(-b+c+\sqrt{cd})}{(b-c+\sqrt{cd})(-a+b+\sqrt{ad})}\right]$$

$$d = a - 2b + c$$

$$a = (\mathbf{r}_3 - \mathbf{r}_1)\cdot(\mathbf{r}_3 - \mathbf{r}_1);\quad b = (\mathbf{r}_3 - \mathbf{r}_1)\cdot(\mathbf{r}_3 - \mathbf{r}_2);\quad c = (\mathbf{r}_3 - \mathbf{r}_2)\cdot(\mathbf{r}_3 - \mathbf{r}_2) \tag{B9}$$

where indexes 1, 2, and 3 denote triangle vertexes, A_p is the triangle area.

Integrals (B9) are now programmed in the script `rwg2.m` and used to calculate the impedance matrix. The easiest way to incorporate (B9) and (B8) into the existing "uniform" algorithm is to replace the corresponding diagonal elements of the Green's function. This is done in the function `impmet.m`. The corresponding Matlab codes are located in the Matlab directory of Appendix B. They are equivalent to the codes of Chapter 2 (scattering) except for the different way of calculating the self-coupling terms of the impedance matrix.

The test of the code has demonstrated minor improvement compared to the case of the uniform 9-point quadrature. For example, if one uses the code of Appendix B, the calculations of Chapter 2 are reproduced almost identically, with the difference in surface current magnitudes less than 1%.

REFERENCES

1. S. M. Rao, D. R. Wilton, and A. W. Glisson. Electromagnetic scattering by surfaces of arbitrary shape. *IEEE Trans. Antennas and Propagation*, 30 (3): 409–418, 1982.
2. T. F. Eibert and V. Hansen. On the calculation of potential integrals for linear source distributions on triangular domains. *IEEE Trans. Antennas and Propagation*. 43 (12): 1499–1502, 1995.

INDEX

Active impedance, *see* Impedance, active
Archimedean spiral antenna, *see* Spiral antenna
Antenna
  aperture, *see* Effective aperture
  arrays, *see* Antenna arrays
  circuit, *see* Antenna circuit
  efficiency, 227
  far-field, *see* Antenna far-field
  feed, *see* Feed model
  impedance, *see* Impedance
  loading, *see* Antenna load
  polarization, *see* Polarization
  radiation mode of, 42–44
  receiving mode of, 11–12, 79–80
  resonance, 158
  transmitting mode of, 39–40
Antenna arrays
  active impedance of, 118
  array factor of, 130–131
  bowtie, 143–145
  broadside, 124–127
  circular, 116–117
  dipole, 115–117
  electronic beam steering of, 124
  end-fire, 124, 128–130, 133–139
  Hansen-Woodyard model, 133, 135–136
  linear, 115–116
  monopole, 116–117
  network equations, 123–124
  patch, 256–257
  pattern multiplication theorem, 130–133
  planar, 256–257
  phased, 114, 139–142
  terminal impedance, *see* Impedance, terminal
  two–element array, 119–122
Antenna circuit
  conjugate matched load, 82
  equivalent circuit, 81, 215
  reflection coefficient, 160, 215
  return loss, 160
  terminal impedance, *see* Impedance, terminal
  transmission coefficient, 215
Antenna far-field
  bandwidth, 152
  beamwidth, 125–126
  cuts
    E-plane, 46, 205
    H-plane, 46, 205
  directivity, 47–48
  directivity patterns, 47–49, 76–79
  effective aperture, 51–52
  far-field distance, 44
  gain, 50–51
  Poynting vector, 45
  radiated field at a point, 44–45
  radiated power, 72, 75
  radiation density, 46
  radiation efficiency, 227
  radiation intensity, 46
  radiation resistance, 74–76
  side lobes, 125
Antenna load
  capacitive load, 230
  lumped elements, 224–225
  modification of impedance matrix, 226
  resistive load, 227–229

Balanis, C. A., 2
Bandwidth, 152
Barycentric triangle subdivision, 18
Beamwidth, 125–126
Bi-directional antenna, 169
Bowtie antenna
  gain, 166, 168
  input impedance, 166–167
  mesh generator, 165–166
  radiation patterns, 169–171
  resonance, 166, 168
  scattering, 28–29
  surface current distribution, 29
  V-shaped, 188
Broadband antenna, 152
Broadside array, 124–127

Capacitive loading, see Antenna load
Capacitor model for dielectric, 248–249
Cavity, 262
Circuit, see Equivalent circuit
Circular array, 116–117
Complementary antennas, 32
Conjugate matched load, 82
Cube, 261
Current radiator, 93
Cylinder, 264

Dipole antenna
  capacitive load, 230
  feed model, 60–62
  gain, 160–162
  input impedance, 65–66, 158–160
  mesh generator, 60
  radiation patterns, 73, 76–77
  resistive load, 227–229
  resonance, 158, 160–162
  resonant length, 158
  return loss, 160–162
  scattering, 25–28
  strip model, 60
  strip model, comparison with NEC solver, 162–165
  surface current distribution, 63–64
  with reflector, 84–86
Dipole array, see Antenna arrays
Directivity, 47–50

Directivity patterns, 48–49, 76–78
Disk, 37

Effective aperture, 51–52
Efficiency of antenna, 227
End-fire array, 124, 128–130, 133–139
E-plane, see Antenna far-field, cuts
Equivalent circuit, 81, 215–216

Far-field, see Antenna far-field
Far-field distance, 44
Feed model
  array, 115, 118
  bottom feed, 69–70
  center feed, 60–62
  coaxial probe feed, 234, 236
  delta gap voltage generator, 61–62
  gap voltage, 11
  feeding/driving edge, 61, 63, 70
  magnetic frill, 237
  voltage excitation vector, 22, 62
Fractal antenna
  gain, 185
  input impedance, 181–182
  input reflection coefficient, 183
  mesh generator, 156–157, 179–180
  radiation patterns, 185
  resonance's, 181, 183–184
  surface current distribution, 184–185
Free-space impedance, see Impedance, free space
Frequency loop, 153–155
Friis transmission formula, 81–82

Gain, 50–51
Gaussian voltage pulse
  duration, 197
  center frequency, 197
  half-power lower frequency, 198
  half-power higher frequency, 198
  half-power bandwidth, 199

Hansen-Woodyard model, 135–137
Helical antenna
  axial mode, 105
  input impedance, 104, 107
  mesh generator, 102, 105
  normal mode, 102
  radiated field at a point, 107

radiation patterns, 103, 106, 108
surface current distribution, 103, 106, 108
H-plane, *see* Antenna far-field, cuts

Impedance
  active, 118
  antenna, 65, 71
  antenna array, 118
  bowtie, *see* Bowtie antenna
  dipole, *see* Dipole antenna
  free space, 24–25
  matrix, *see* Impedance matrix
  monopole, *see* Monopole antenna
  mismatch, 213, 215–216
  mutual impedance, 123
  self-impedance, 123
  spiral, *see* Spiral antenna
  terminal, 118–120, 123–124
  ultrawideband antenna, 203
Impedance matrix
  antenna loading, 226
  calculation, 265
  filling time, 8
  integration, 18
  maximum size, 8, 144–145
  metal structure, 19–21
  metal-dielectric structure, 249
  RWG edge elements, 5–7, 16–17
  self-coupling terms, 266
Impulse antenna, 193–194
Input impedance, *see* Impedance
Input reactance
  capacitive, 95
  inductive, 95
Input resistance
  loss component, 227
  radiation component, 74–76
Isotropic antenna, 48

Kraus, J. D., 115, 119

Linear array, 115–116
Load, *see* Antenna load
Loop antenna
  axial mode, 100–102
  electrically large, 93
  electrically small, 93
  input impedance, 95, 102

  mesh generator, 90–91
  radiated field at a point, 98
  radiation patterns, 96–97, 99, 101
  surface current distribution, 94
Lumped elements, *see* Antenna load

Matlab
  antenna related packages, 1
  compiler, 33–34
  disabling Java virtual machine, 145
  function(s), *see* Matlab functions
  execution times, 8
  loop(s), 7. *See also* Parameter/Frequency loop
  matrix solver(s), 22–23
  mesh, *see* Mesh generator(s)
  mouse input, 67, 237
  PDE toolbox, 14–16, 28, 30–31, 199, 245
  scripts, converting to functions, 33
  under LINUX, 8, 34
  under Windows, 8
Matlab functions
  `angle`, 207
  `delaunay`, 165
  `delaunay3`, 14
  Gaussian elimination, 23
  `ginput`, 67
  `gmres`, 23
  `inv`, 23
  `rotate3d`, 16
  `polar`, 48
  `subplot`, 48
  `view`, 13
  `unwrap`, 207
Mesh(es)
  antenna, *see* Mesh generator(s)
  cloning of, 115
  generator(s), *see* Mesh generator(s)
  structured, 15
  unstructured, 15
Mesh generator(s)
  array of bowties, 141
  array of dipoles, circular, 116–117
  array of dipoles, linear, 115–116
  array of monopoles, 116–117
  bowtie
    Matlab script, 165–166
    V-shaped, 188

cavity, 262
cube, 261
cylinder, 264
dipole, 60
dipole with reflector, 85–86
disk, 37
fractal antenna, 156–157, 179–180
helical antenna, 102, 105
loop antenna, 90–91
metal lattice, 260, 261, 263
mouse input, 67, 237
overview, 4
patch antenna, 237–239, 245–246, 250–251
patch array, 256–257
PDE toolbox mesh generator, 14–16, 28, 30–31, 199, 245
plate
   bent, 84–85
   Matlab script, 67, 259
reflector, 84–86
slot, 30
slot antenna of Time Domain, Co., 199
sphere, 264
spiral, 155, 172–173
strip, 60
volume metal mesh, 260–261
Method of Moments (MoM)
   equations of,
      scattering, 21–22
      radiation, 62
      metal-dielectric, 249
   for pulse antenna, 195
   impedance matrix, *see* Impedance matrix
   in time domain, 195. *See also* Time-domain analysis
MoM, *see* Method of Moments
Monopole antenna
   feed model, 70
   gain, 74
   input impedance, 71
   mesh generator, 67–70
   radiated field at a point, 72
   radiation patterns, 72–74
   resistive load, 232
   surface current distribution, 69

Monopole array, *see* Antenna arrays
Mouse input, *see* Matlab
Multiband antenna, *see* Fractal antenna

Omnidirectional antenna, 50

Pattern multiplication theorem, 130–133
Parameter loop
   versus element spacing for broadside array, 122–123
   versus phase shift for end-fire array, 133–135
Patch antenna
   array, 256–257
   dielectric model, 248–249
   feed model, 234, 236–237
   input impedance, 240
   mesh generator, 237–239, 245–246, 250–251
   radiation patterns, 242, 244
   resonance, 241, 244, 247, 252
   return loss, 240, 244, 251–252
Patch array, 256–257
Phase, *see* Matlab functions, unwrap
Phased array, 114, 139–142
Planar array, 256–257
Plane wave, 22
Plane wave approximation, 44
Plate
   bent, 84–85
   mesh using PDE toolbox, 15
   mesh using plate.m, 67, 259
   scattering, 24–25
Polarization
   circular, 107
   dipole versus slot, 32
   linear, 22, 27
Power gain, *see* Gain
Power resonance's, 158, 160
Poynting vector, 45
Progressive phase shift, 128. *See also* Phased array
Pulse
   bandwidth, *see* Gaussian voltage pulse
   center frequency, *see* Gaussian voltage pulse
   fidelity, 209

Radiated field at a point, 44–45
Radiated power, 72
Radiation
  density, 46
  efficiency, *see* Efficiency of antenna
  intensity, 46
  resistance, 74–76
Radiation of surface currents, 42–44
Receiving antenna, 11–12, 79–80, 209–210
Reflection coefficient, 160, 215. *See also* Impedance
Reflector, 84–86
Resistive loading, *see* Antenna load
Resonant length, 158
Return loss, 160. *See also* Impedance
RWG edge elements, 5–7, 16–17

Scattering
  bowtie, 29
  dipole, 27
  incident field, 22, 26, 28, 30
  plate, 24
  slot, 31–32
Side lobes, 125
Slot, *see* Mesh generator(s)
Slot antenna
  input impedance for different feed models, 201
  mesh, 29–30
  polarization, 31–32
  scattering, 29–32
  ultra-wideband, 193–194
Sphere, 264
Spiral antenna
  active region, 176
  conical, 190
  gain, 177
  input impedance, 172–176
  mesh generator, 155, 172–173
  radiation patterns, 178–179
  resonance, 174
  surface current distribution, 176

Strip, *see* Mesh generator(s)
Strip mesh, 60
Surface electric currents
  density, 21
  radiation of, 42–44
  visualization, 23–24
  independent of frequency, 223

Terminal impedance, *see* Impedance, terminal
Time-domain analysis
  antenna-to-antenna transfer function, 209–211
  antenna-to-free-space transfer function, 207–209
  discrete Fourier transform, 211–213
  MoM for time domain, 195
Transverse electromagnetic (TEM) waves, 44
Transfer function, *see* Time-domain analysis
Transmission coefficient, 215
Transmitting antenna, 39–40

Ultra-wideband antenna, 152, 193–194
Ultra-wideband antenna of Time Domain. Co.
  feed model, 199–200
  input impedance, 201, 203
  mesh generator, 199–200
  radiated field at a point, 208
  radiation patterns, 205–206
  surface current distribution, 202

Voltage radiator, 93
Volume metal mesh, 260–261

Wire antenna
  strip model of a wire, 7, 60
  radius of equivalent wire, 60

Yagi-Uda antenna, 138

CUSTOMER NOTE: IF THIS BOOK IS ACCOMPANIED BY SOFTWARE, PLEASE READ THE FOLLOWING BEFORE OPENING THE PACKAGE.

This software contains files to help you utilize the models described in the accompanying book. By opening the package, you are agreeing to be bound by the following agreement:

This software product is protected by copyright and all rights are reserved by the author and John Wiley & Sons, Inc. You are licensed to use this software on a single computer. Copying the software to another medium or format for use on a single computer does not violate the U.S. Copyright Law. Copying the software for any other purpose is a violation of the U.S. Copyright Law.

This software product is sold as is without warranty of any kind, either express or implied, including but not limited to the implied warranty of merchantability and fitness for a particular purpose. Neither Wiley nor its dealers or distributors assumes any liability of any alleged or actual damages arising from the use of or the inability to use this software. (Some states do not allow the exclusion of implied warranties, so the exclusion may not apply to you.)

WILEY